# Microbial Communication

Sarangam Majumdar • Sisir Roy

# Microbial Communication

## Mathematical Modeling, Synthetic Biology and the Role of Noise

Springer

Sarangam Majumdar
Dipartimento di Ingegneria e Scienze
dell'Informazione e Matematica
Università degli Studi dell'Aquila
L'Aquila, Italy

Sisir Roy
IISC Campus
National Institute of Advanced Studies
Bangalore, Karnataka, India

ISBN 978-981-15-7416-0 ISBN 978-981-15-7417-7 (eBook)
https://doi.org/10.1007/978-981-15-7417-7



This Springer imprint is published by the registered company Springer Nature Singapore Pte Ltd.
The registered company address is: 152 Beach Road, #21-01/04 Gateway East, Singapore 189721, Singapore

*To our family and friends*

# Preface

Bacteria are microorganisms, which communicate with each other using three different mechanisms. Historically, the chemical signalling mechanism was first discovered and extensively discussed by scientists. Chemical molecules have been identified for this mechanism and they are technically known as autoinducers. Communication through sound is discussed by the end of last century but not so well mentioned in the literature. The most recent mechanism was discovered a couple of years before by the researchers from the USA and the European Union known as communication through electrical signalling. These three types of communication are used by the bacteria to make their own survival master plan in a complex environmental condition. The chemical signalling mediated communication process depends on the number density of the bacterial cell. This dependence of number density is commonly known as a quorum sensing mechanism. The study of the mechanisms for communication among the bacteria needs an interdisciplinary approach. The main object of this book is to analyse various types of bacterial communications using mathematical modelling.

The exploration of this book starts with the analysis of basic concepts of bacterial communication and reviewing the significant experimental observations. Then, we move onto mathematical models of bacterial communication mechanism and quorum sensing regulated processes such as biofilm formation, gene expression, bioluminescence, swarming, virulence and so on, to understand the underlying process of the complex biochemical communication systems and continue with the evolutionary type of models, pattern formation and therapy related models (quorum quenching). Then we summarize the important observations of mathematical models and experimental results. The crucial role of noise in these biological communication systems and the bacterial ion-channel mediated communication process, bacterial reconciliation and time sharing behaviour are discussed.

Later, we focus on artificial bacterial cells and communication between artificial and natural cells, as well as we construct fundamental technological platforms for characterization of communication circuits within the view of synthetic biology. The various aspects of synthetic biological approaches (related with bacterial communication) such as genetic toggle switch, bistable gene regulatory network, transcriptional repressor system, pattern formation, synthetic bacterial population control circuit, predator–prey synthetic system, synchronized quorum of genetic clocks, the role of noise in synthetic biology and stochastic Turing patterns in bacterial population have been discussed.

The motivation of writing this book is to not only unfold the mathematical aspects of a bacterial communication system but also present in such a manner so that non-specialist reader can feel the taste of the field. Our main target readers are undergraduate students, graduate students, young researchers as well as experts in science and technology who will be interested in general. It will be also useful for the readers interested especially working in mathematical biology and the interface of biology and physics. We hope the readers of this book will find some interesting and fundamental insight into the bacterial communication processes.

L'Aquila, Italy Sarangam Majumdar
Bangalore, India Sisir Roy
February, 2020

# Acknowledgements

The authors are extremely grateful to all their collaborators and family members for their supports and valuable suggestions. One of the authors (Prof. Sisir Roy) is indebted to Homi Bhabha Fellowships Council, Mumbai for their financial assistance to write this book. The authors acknowledge the tireless effort of Dr. Malabika Roy who read the entire manuscript and suggested some changes. Finally, we express our deepest gratitude to the editors, reviewers and the production team of the Springer Nature for accepting our proposal and their entire effort to publish this book.

# Contents

# About the Authors

**Sarangam Majumdar** is a mathematician. He holds a Bachelor of Science and Master of Science in Mathematics from the University of Calcutta (2010) and National Institute of Technology, Rourkela (2012), respectively. He has worked with Prof. Sisir Roy, Indian Statistical Institute, Kolkata; Prof. Rodolfo Llinas, New York University School of Medicine, USA; Prof. Dr. Suman Datta, University of Calcutta; Prof. S. Chakraverty, National Institute of Technology, Rourkela; Dr. Sukla Pal, University of Otago, New Zealand; Prof. Rodolfo Repetto, Dipartimento di Ingegneria delle Costruzioni, dell'Ambiente e del Territorio, Università degli Studi di Genova, Italy; Prof. Nicola Guglielmi, Università degli Studi di L'Aquila; and many more. His research interests include mathematical biology, system biology, mathematical modeling, numerical analysis, scientific computing, and nonlinear dynamics and chaos, and over the past 8 years, he has focused on bacterial communication systems (quorum sensing and electrical communication). He has published several papers in various peer-reviewed, international journals and book chapters (in Springer). He is also a reviewer for a number of international journals.

**Prof. Sisir Roy** is a theoretical physicist currently working as a Visiting Professor and Senior Homi Bhabha Fellow, National Institute of Advanced Studies, IISc Campus, Bangalore. Previously, he worked as a Professor at the Physics and Applied Mathematics Unit, Indian Statistical Institute, Kolkata, from 1980 to 2014 and as a Visiting Scientist at the Institute Henri Poincare, Paris, France, from 1986 to 1987. He has also worked as a Distinguished Visiting Professor at numerous US and European universities. He has collaborated with Prof. J.P. Vigier, France; Prof. Bo Lehnert, Alfven Laboratory, Sweden; Prof. Menas Kafatos, George Mason University; Prof. Ralph Abraham, University of California; and Prof. Rodolfo Llinas, New York University School of Medicine, NY, USA. His main interests include the foundations of quantum theory, theoretical astrophysics, quantum biology, brain function modeling, and higher-order cognitive activities. He has published more than 200 papers in various peer-reviewed, international journals, 12 monographs and edited volumes published by Kluwer Academic Publishers, World Scientific, Singapore, Springer, and others. He is currently a member of the editorial board of various international journals. His recent books include Sisir Roy (2016), Decision Making and Modeling in Cognitive Science, Springer.

# Introduction

# 1

**Abstract**

Bacteria are unicellular microorganism, which can communicate through three different ways such as chemical signalling mechanisms, acoustic wave process and electrical signalling mechanism for their survival in distinct environmental conditions. Chemical signalling process is formally known as quorum sensing, which is extensively studied by many microbiologists in the last three decades and reported that this microbial communication process regulated several others biochemical phenomena such as bioluminescence, gene expression, biofilm formation, swarming, virulence and many more. In this chapter, we introduce the overview of microbial communication system and its indispensability as an interdisciplinary subject.

## 1.1 Bacteria and Bacterial Communication

Bacteria are one of the domains in the three domains of life, which plays an important role in our everyday life. This single cellular microscopic organisms are possibly the first indication of life in earth. Antonie Van Leeuwenhoek reported about the concept of microbes (including bacteria) with his self-designed microscope, in a letter, protozoa [1]. There, he also pointed out about a typical characteristic regarding bacteria, i.e. “these species live in a large environment”. Before twentieth century, many scientists ignored or even doubted about the clarity and resolution of Leeuwenhoek’s microscope, but, during the last century, the whole extent of works regarding the characteristics of bacteria established Leeuwenhoek’s finding of bacterial cells, having resolution less than one micro meter [2]. So, an era began with his remarkable observation. In the last 350 year several ground breaking works were published, which have enriched the field of microbiology (Fig. 1.1).


S. Majumdar, S. Roy, *Microbial Communication*,
https://doi.org/10.1007/978-981-15-7417-7_1

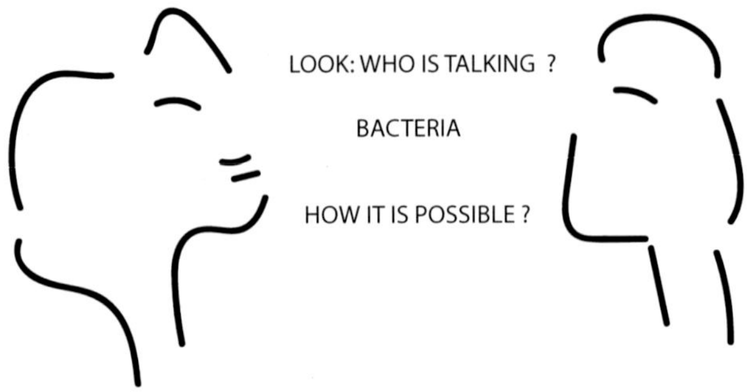


**Fig. 1.1** Pictorial representations of the main theme of the book

### 1.1.1 History and Quorum Sensing

Harvey [3] studied luminescent bacteria in 1952 and reported that growth and light production are closely related. In the late 1960s, Kempner and Hanson [4] performed several experiments on the light production by photobacterium *fischeri* (later known as *V. fischeri*) and suggested that the inhibitor binds with luciferase molecules. Nealson and his collaborators [5] put forwarded that the autoinduction of luminescence happens on the transcriptional level. Further, Eberhard [6] explained that the luciferase synthesis is repressed in fresh medium and it is induced after a certain time. Before 1994, this biochemical process is known as autoinduction. But a confusion occurred with the term autoinducers (diffusible molecules involved in this process). To resolve this confusion Steven Winans called this complex biochemical process as "quorum sensing" [7] and, finally, this bacterial communication process is formally known as quorum sensing mechanism.

In this communication mechanism, bacterial cells emit and receive chemical signalling molecules called quorum sensing molecules (or autoinducers) to regulate the whole system. Gram-negative bacteria utilize acylated homoserine lactones as autoinducers, whereas gram-positive bacteria utilize modified oligopeptides. Autoinducer plays a central role to direct the communication process. Quorum sensing can be defined as a gathering of high autoinducer concentrations at high bacterial population densities. In such a situation, concerning the response associated with a population-wide alteration of gene expression, these bacteria allow the community to coordinate their behaviour in a way, quite similar to cells in a multicellular organism. This constitutes of various steps, synthesis of autoinducer (signal) molecules, diffusion of autoinducer (signal) across the cell membrane, complex formation between autoinducers (signal), the regulatory protein and manufacture of the target protein (see details in Chaps. 2 and 9). Presence of Quorum sensing mechanism can also be observed in fungi and other insects (i.e. ants and honey bee). This bacterial collective behaviour regulated several processes which include biofilm formation, virulence, gene expression, swarming and many more [8–11]. Such type of communication is usually known as microbial communication as mentioned in the title of the book. We are trying to gain a fundamental insight of bacterial communication systems throughout this book.

### 1.1.2 Bacterial Electrical Communication

Besides chemical communication process, bacteria have the ability to communicate via electrical signalling mechanism. Bacterial ion-channels conduct electrical communication process within bacterial biofilm communities. This whole communication process is mediated by the potassium wave [12]. It has already been confirmed that metabolic codependence between peripheral and interior bacterial cells of biofilms evolved oscillatory behaviour within biofilms. As a consequence biofilm growth is halted periodically [13]. Distant motile cells are also attracted towards biofilm through electrical signalling [14, 15]. Later, we observe that distant biofilms are coupled through the electrical signalling (see details in Chap. 12). Furthermore, time sharing behaviour is discovered in limiting nutrient condition [16, 17].

### 1.1.3 Bacterial Acoustic Communication

Bacterial cells are also efficient to communicate via sound wave. Some studies suggested that *B. carboniphilus* growth is controlled by sound wave (under stress condition). This growth stimulation is sonic in nature. Sensitive microphone can also detect sound wave from gram-positive *B. subtilis* bacteria [18–20].

## 1.2 Mathematical Modelling

When any kind of phenomena (i.e. physical, biological, chemical, etc.) can be presented with the proper application of mathematical formalism, then the model, thus developed, is called a mathematical model of that particular phenomena. Normally, different kinds of mathematical approaches are applied with the aim of understanding the nature of these phenomena and then that of controlling the systems so that this procedure guides the researchers towards a direction, leading ultimately to the fruitful solutions of the concerned problems. To build up such a model, it is crucially important to understand the complexity involved in the cases concerned followed by the proper applications of those models which ultimately leads us to explain different facts. In order to gain this achievement, the next step consists of the application of proper, suitable, mathematical equations (for example, ordinary differential equations, partial differential equations, those of linear/nonlinear origin, etc. as per the requirement, for finding the proper solutions). A fundamental point to be noted at this point is that the whole mathematical modelling is based on the current understanding of the system to be analysed mathematically, but with the help of the particular model of interest, best suitable for the expected information to be achieved. In most of the cases, the model is based on certain assumptions, considered as vital in making a good, viable mathematical mode for the phenomena to be explained. In general, models are basically of two types: deterministic or stochastic models. Based on the consideration of several factors, for example, production of common good, spatial constrains and population dynamics, to name a few, the problem can be simulated, basically utilizing either deterministic modelling approach when average behaviour of the population is sufficient to study quorum sensing or stochastic modelling approach, i.e. whenever variation in quorum sensing among the individuals in a population occurs. Both of them are expected to be able of expressing the natural phenomenon by analysing the associated systems.

In this monograph, we did put our best efforts in dealing a few types of mathematical models, which have already been established. For example, mathematical models of quorum sensing, molecular mechanisms (see details in Chap. 3), therapy related models (see details in Chap. 10), evolutionary models (see details in Chap. 7), biofilm models (see details in Chap. 4), swarming models (see details in Chap. 5), virulence models (see detail in Chap. 6), pattern formation (see details in Chap. 8) are to name a few among them.

## 1.3 Synthetic Biology

Synthetic biology is considered as a new branch of science where scientists are redesigned existing biological systems using the design principle of new biological entities such as genetic circuits, enzymes and cells. This branch of biology, a rapidly growing field of research being an interdisciplinary branch of science, is involved in re-engineering organisms for beneficial purposes, applying engineering concepts and principles to have the new capabilities with the purpose of the sensible engineering of living system, such as yeast and bacteria. Scientists and companies all around the world are involved in a dedicated manner in exploiting the power of nature in order to find out the proper solutions for the tough, almost non-negotiable up till date, of the problems especially connected to the medicine, agriculture and energy saving manufacturing.

In this arena, basic science collaborates with that of engineering to delineate an artificial biological system for understanding that of complex biological origin and apply it in industries and medical sciences [21–23]. Here, living cells are used as the substrate for implementing human-defined computations. Many current implementations have already been in use, for example, the cellular computing which are constructed on the "genetic circuit" metaphor, being an approximation of the operation of silicon-base. Although this visionary mapping has been comparatively successful, the group led by Angel Goni-Moreno [24] argued regarding the fundamental constraints, limiting the varieties of estimation that may be engineered within the cell at the cost of exploiting the wealthy and diverse functionality, obtainable in natural living systems (Fig. 1.2).

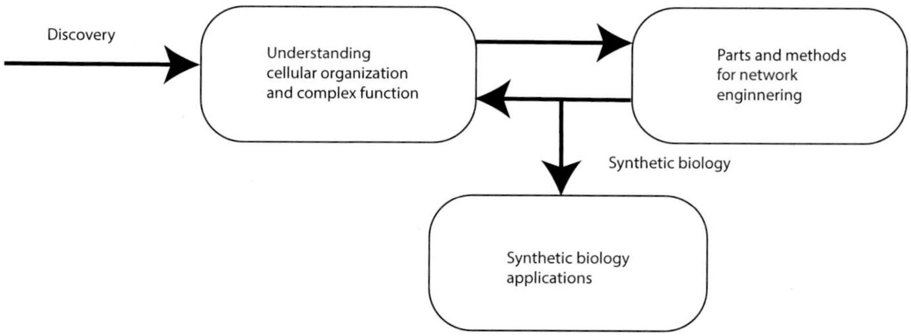


**Fig. 1.2** Pictorial representations of the engineering approach and complementarity of discovery for better understanding of biological complex systems. Synthetic biology reveals design principle and organization which improve our understanding of living systems with useful application

According to his opinion, this branch of biological engineering science possesses a great potentiality with the ability for serving the humankind in various aspects. These might be of various origins, for example, design and manufacturing of new biological parts, devices and systems together with the re-engineering of existing, natural biological systems. Or, to be more specific, for useful purposes involved in this branch of science, the basic need is to redesign organisms by engineering theme so that it would be possible in possessing new abilities, by harnessing the power of nature, for example, solving typical problems in medicine, manufacturing as well as agriculture. That is why, in some ways, the synthetic biology could find itself quite similar to another approach, i.e. "genome editing", as, in both of these processes, altering an organism's genetic code is involved. However, following a school of researchers, a distinction between these two processes could be drawn because distinct differences in the modality of the change exist. In synthetic biology, the scientists, involved in the research, primarily engaged themselves in stitching together long stretches of DNA, followed by inserting them in an organism's genome. These synthesized pieces of DNA could be either completely novel in nature or in the genes, found in other organisms or entirely, they could be novel. For example, in genome editing, scientists, specifically, try to avail those tools, trying to make changes, as small as possible, compared to the organism's own DNA. Genome editing tools can also be used to delete or add small stretches of DNA in the genome. Thus the pledge of synthetic biology reclines in its potentiality in providing new substrates for production, computation, pollution control and medical diagnosis, and to exploit the "wetware" within the living cell for human-defined purposes [24].

But the prime challenge behind this fascinating branch is to realize general, scalable strategies that certify creation of increasingly complex gene circuits with dependable performance, as well as to construct fundamental and significant technological platforms for the quantitative circuit characterization [25–27]. This branch of biology becomes widespread in the beginning of the twenty-first century with very notable publications of Collins and his co-workers (synthetic genetic toggle switch, necessary condition for bistability) [28], Elowitz and Leibler (design principle and construction of synthetic network in bacteria) [29]. Following this track, several researches have carried out their research works , for example, cell -to -cell communication systems and communication between artificial and natural cells, pattern formation, synthetic cooperation, predator–prey synthetic system, dynamical quorum sensing, synchronized quorum of genetic clocks, role of noise in synthetic biology and stochastic Turing test (see details in Chap. 13), to name a few, which shed a new light in the field related to bacterial communication.

## 1.4 Noise in Biology

After following a revolutionary research, Monod [30] opined "Living world is shaped by the interplay of deterministic laws and randomness". If any scientist wants to uncover any kind of natural phenomenon, then he or she has to go through mysterious noisy environment. Biological science deals with the living systems which are obviously non-equilibrium, and so it is not astonishing that noise plays a vital role in many biological processes [31]. Researchers have started in detecting more and more evidence that noise is not always harmful for a biological function. The evolutionary process leads us in tuning the systems so that they can take an advantage of natural stochastic fluctuations. The functional roles of noise in biological processes possess a great diversity by permeating in all levels, starting with standard, entropy-increasing effects, then, playing occasionally, more surprising constructive roles, for example, accelerating the speed of evolution by providing selective advantage in dynamic environments. This gifted phenomena of surprising nature, also contributes in increasing intracellular transport of bio molecules, thereby increasing information capacity of signalling pathways, producing random mutations, diversifying phenotypes in isogenic populations, limiting information capacity of signalling relays, or, in short, starting from the most basic molecular, sub-cellular processes up to the dynamics of tissues, organs, organisms, and populations. Thus this phenomenon can be termed as one of the crucial signals in the everyday world, because no one can deny its presence.

In engineering field (i.e. communication science, electronics computation, etc.), people usually consider noise as an unwanted variation and try to optimize. But, though in general sense, noise is considered to have a destructive role, for example, especially, in case of signal analysis (science and technology), it has already been an established phenomenon that the input noise refines detectability and transduction of signal in nonlinear systems, usually called as Stochastic Resonance. Thus, the discovery of Stochastic Resonance brought a paradigm shift to the perception (i.e. destructive role of noise) which has been found in biological systems and considered as a central parameter in system function [32]. Not only that, noise plays a fundamental role in developmental biology, cell communication, cellular decision making, ion-channels, brain function, gene expression and many more. However, in this short review, we, with our earnest efforts, try to present a few of the recent progress in understanding mechanisms and effects of fluctuations in biological systems of different scales, for example, the role of noise in quorum sensing circuits (see details in Chap. 11), synthetic systems (see details in Chap. 14) as well as biological noise verses noise in science and technology (see details in Chap. 15).

## References


1. Leewenhoeck A (1677) Observation, communicated to the publisher by Mr. Antony van Leewenhoeck, in a Dutch letter of the 9 October 1676 here English'd: concerning little animals by him observed in rain-well-sea and snow water; as also in water wherein pepper had lain infused. Philos Trans 12:821–831. https://doi.org/10.1098/rstl.1677.0003
2. Lane N (2015) The unseen world: reflections on Leeuwenhoek (1677) concerning little animals. Philos Trans R Soc B 370(1666):20140344. https://doi.org/10.1098/rstb.2014.0344
3. Harvey EN (1952) Bioluminescence. Academic Press, New York
4. Kempner ES, Hanson FE (1968) Aspects of light production by *Photobacterium fischeri*. J Bacteriol 95:975–979
5. Nealson KH, Platt T, Hastings JW (1970) Cellular control of the synthesis and activity of the bacterial luminescent system. J Bacteriol 104:313–322
6. Eberhard A (1972) Inhibition and activation of bacterial luciferase synthesis. J Bacteriol 109:1101–1105
7. Turovskiy Y, Kashtanov D, Paskhover B, Chikindas ML (2007) Quorum sensing: fact, fiction, and everything in between. Adv Appl Microbiol 62:191–234
8. Miller MB, Bassler BL (2001) Quorum sensing in bacteria. Ann. Rev. Microbiol. 55(1):165–199
9. Fuqua WC, Winans SC, Greenberg EP (1994) Quorum sensing in bacteria: the LuxR–LuxI family of cell density-responsive transcriptional regulators. J Bacteriol 176(2):269
10. Waters CM, Bassler BL (2005) Quorum sensing: cell-to-cell communication in bacteria. Annu Rev Cell Dev Biol 21:319–346
11. Majumdar S, Mondal S (2016) Conversation game: talking bacteria. J Cell Commun Signal 10(4):331–335
12. Prindle A, Liu J, Asally M, Ly S, Ojalvo GJ, Süel MG (2015) Ion channels enable electrical communication in bacterial communities. Nature 527(7576):59–63
13. Liu J, Prindle A, Humphries J, Sagarra MG, Asally M, Dong-yeon LD, Ly S, Ojalvo GJ, Süel MG (2015) Metabolic co-dependence gives rise to collective oscillations within biofilms. Nature 523(7562):550–554
14. Humphries J, Xiong L, Liu J, Prindle A, Yuan F, Arjes AH, Tsimring L, Süel MG (2017) Species-independent attraction to biofilms through electrical signaling. Cell 168(1):200–209
15. Majumdar S, Pal S (2017) Cross-species communication in bacterial world. J Cell Commun Signal 11(2):187–190
16. Liu J, Corral MR, Prindle A, Dong-yeon LD, Larkin J, Sagarra GM, Ojalvo GJ, Süel MG (2017) Coupling between distant biofilms and emergence of nutrient time-sharing. Science 356(6338):638–642
17. Majumdar S, Pal S (2017) Bacterial intelligence: imitation games, time-sharing, and long-range quantum coherence. J Cell Commun Signal 11(3):281–284
18. Matsuhashi M, Pankrushina AN, Takeuchi S, Ohshima H, Miyoi H, Endoh K, Murayama K, Watanabe H, Endo S, Tobi M, Mano Y (1998) Production of sound waves by bacterial cells and the response of bacterial cells to sound. J General Appl Microbiol 44(1):49–55
19. Scholkmann F, Fels D, Cifra M (2013) Non-chemical and non-contact cell-to-cell communication: a short review. Am J Transl Res 5(6):586–593
20. Reguera G (2011) When microbial conversations get physical. Trends Microbiol 19(3):105–113

21. Nandagopal N, Elowitz MB (2011) Synthetic biology: integrated gene circuits. Science 333(6047):1244–1248
22. Khalil AS, Collins JJ (2010) Synthetic biology: applications come of age. Nat Rev Genet 11(5):367–379
23. Bashor CJ, Horwitz AA, Peisajovich SG, Lim WA (2010) Rewiring cells: synthetic biology as a tool to interrogate the organizational principles of living systems. Annu Rev Biophys 39:515–537
24. Grozinger L, Amos M, Gorochowski TE, Carbonell P, Oyarzún DA, Stoof R, Fellermann H, Zuliani P, Tas H, Goñi-Moreno A (2019) Pathways to cellular supremacy in biocomputing. Nat Commun 10(1):1–11
25. Simpson ML (2004) Rewiring the cell: synthetic biology moves towards higher functional complexity. Trends Biotechnol 22(11):555–557
26. Endy D (2005) Foundations for engineering biology. Nature 438(7067):449–453
27. Marguet P, Balagadde F, Tan C, You L (2007) Biology by design: reduction and synthesis of cellular components and behaviour. J R Soc Interface 4(15):607–623
28. Gardner TS, Cantor CR, Collins JJ (2000) Construction of a genetic toggle switch in *Escherichia coli*. Nature 403(6767):339–342
29. Elowitz MB, Leibler S (2000) A synthetic oscillatory network of transcriptional regulators. Nature 403(6767):335–338
30. Monod J (1971) Chance and necessity: an essay on the natural philosophy of modern biology. Alfred A. Knopf, New York. ISBN 0-394-46615-2
31. Tsimring LS (2014) Noise in biology. Rep Progr Phys. 77(2):026601
32. Roy S, Llinás R (2012) The role of noise in brain function. In: Science: image in action, pp 34–44

# 2 Talking About Talking Microbes

**Abstract**

Bacterial quorum sensing mechanism is considered as the gene expression regulator in response to fluctuations in bacterial cell population density. This communication process is controlled by autoinducers. So bacteria can talk to each other using autoinducers. We introduce bacterial talking mechanism or communication process in this chapter. We briefly discuss quorum sensing process in cases of different bacteria such as LuxI/ LuxR type quorum sensing, LasI/LasR- RhlI/RhlR system, TraI/TraR system, ExpI/ExpR-CarI/CarR system, ComD/ComE system, ComP/ComA system, AgrC/AgrA system and LuxS family (interspecies communication). Here, we study the communication among the bacteria through chemical signalling only.

## 2.1 Bacterial Quorum Sensing Mechanism

Bacteria secrete molecules which are used for their communication with other surrounding bacteria (interspecies and intraspecies). This small secreted diffusible molecule is a key controller of the communication mechanism which is formally known as autoinducer or quorum sensing molecule (QSM) or chemical signalling molecule. Bacteria receive these chemical signals from other bacteria with the purpose of coordinating a collective behaviour. Bacteria emit and receive small chemical signal in order to extend in concentration as a function of bacterial cell number density. An important factor to be mentioned is that when bacteria continue to emit autoinducers in the environment, then the external concentration of the autoinducers is directly proportionate to the cell population density, making bacteria aware of the threshold concentration of the autoinducers as a result of which, gene expression starts altering [1–3]. Thus, we can say it as bacterial quorum sensing mechanism or chemical signalling mechanism. Bacterial communication systems regulate variety of physiological activity, which include biofilm formation, motility,



S. Majumdar, S. Roy, *Microbial Communication*,
https://doi.org/10.1007/978-981-15-7417-7_2

**Fig. 2.1** Quorum sensing: Bacteria emit autoinducers at low cell density, but they are not able to communicate with the surrounding bacteria. Bacteria emit and receive autoinducers at high cell density and the autoinducers concentration achieves a threshold. Quorum sensing begins at that point of time. This bacterial collective behaviour is a density dependent phenomenon

symbiosis, sporulation, virulence, conjugation, competence, antibiotic production. Quorum sensing was first observed in marine bacterium called *Vibrio fischeri*, which can be found as living microorganism as well as a symbiont in the light producing organ of an animal host (i.e. Hawaiian bobtail squid). *V. fischeri* is non-luminescent at low density, when the cell population grows up at a certain level and autoinducers concentration reaches a threshold, a coordination change is initiated. At that point of time, gene expression takes place and generates the enzyme luciferase, which leads to bioluminescence [2]. So, it is very much understandable that bacteria are talking to each other via small molecule as a collective behaviour which we call quorum sensing (Fig. 2.1).

Gram-negative bacteria use N-acyl homoserine lactones (HSL), fatty acid methyl esters, alkyl quinolones as autoinducers (chemical signalling molecules) and gram-positive bacteria use oligo peptides for conversation. Here we track some quorum sensing bacteria with their features in Table 2.1.

## 2.2 Quorum Sensing in Gram-Negative Bacteria

In the last few decades, several gram-negative bacteria are identified, which communicate using chemical signalling molecules or autoinducers (Fig. 2.2). Gram-negative bacterial communication contains at least two homologues regulatory proteins, known as LuxI and LuxR. Biosynthesis of autoinducers (specific acylated homoserine lactone) is controlled by LuxI link proteins and the autoinducers concentration elevates with rise of cell population density. Thereafter, LuxR link protein binds with the autoinducers (specific acylated homoserine lactone) and reaches the threshold concentration. Finally, target gene transcription is activated by

**Table 2.1** List of gram-negative quorum sensing bacteria with chemical signalling molecules, regulatory proteins and phenotypes

| Organism | Chemical signalling molecules | Regulatory proteins | Phenotypes |
|---|---|---|---|
| *Agrobacterium tumefaciens* | 3-Oxo-$C_8$-HSL | TraI/TraR | Ti plasmid conjugation |
| *Aeromonas hydrophila* | $C_4$-HSL | AhyI/AhyR | Exoprotease production |
| *Aeromonas salmonicida* | $C_4$-HSL | AsaI/AsaR | Extracellular protease |
| *Burkholderia cepacia* | $C_8$-HSL | CepI/R | Protease, siderophores |
| *Chromobacterium violaceum* | $C_6$-HSL | CviI/CviR | Exoenzymes, antibiotics, cyanide, violacein |
| *Erwinia chrysanthemi* | 3-Oxo-$C_6$-HSL $C_6$-HSL | ExpI/ExpR | Pectate lyases |
| *Erwinia stewartii* | 3-Oxo-$C_6$-HSL | EsaI/EsaR | Exopolysaccharide, virulence factors |
| *Enterobacter agglomerans* | 3-Oxo-$C_6$-HSL | EagI/EagR | – |
| *Escherichia coli* | – | –/SdiA | Cell division, attachment and effacing lesion formation |
| *Erwinia carotovora* subsp. *carotovora* | 3-Oxo-$C_6$-HSL | ExpI/ExpR CarI/CarR | Exoenzymes Carbapenem antibiotics |
| *Pseudomonas aeruginosa* | 3-Oxo-$C_{12}$-HSL $C_4$-HSL | LasI/LasR RhlI/RhlR | Biofilm formation, multiple extracellular enzymes, Xcp, RhlR secondary metabolites, RpoS |
| *Pseudomonas aureofaciens* | $C_6$-HSL | PhzI/PhzR | Phenazine antibiotics |
| *Pseudomonas syringae* | 3-Oxo-$C_6$-HSL | AhlI/AhlR | Epiphytic fitness, cell aggregation |
| *Pseudomonas chlororaphis* | $C_6$-HSL | PhzI/PhzR | Phenazine-1-carboxamide biosynthesis |
| *Pseudomonas putida* | 3-Oxo-$C_{12}$-HSL | PpuI/PpuR | Biofilm development |
| *Pseudomonas fluorescens* | Long acyl-chain-HSL | MpuI/MpuR | Mupirocin biosynthesis |
| *Rhizobium leguminosarum* | $C_6$-HSL | RhiI/RaiR | RhiABC rhizosphere-expressed genes, nodulation |
| *Rhizobium etli* | – | RaiI/RaiR | Restriction of number of nitrogen fixing nodules |
| *Ralstonia solanacearum* | $C_8$-HSL | SolI/SolR | – |

(continued)

**Table 2.1** (continued)

| Organism | Chemical signalling molecules | Regulatory proteins | Phenotypes |
|---|---|---|---|
| *Rhodobacter sphaeroides* | 7-*cis*-$C_{14}$-HSL | CerI/CerR | Dispersal from bacterial aggregates |
| *Serratia liquefaciens* | $C_4$-HSL | SwrI/SwrR | Extracellular protease, swarming |
| *Salmonella typhimurium* | – | –/SdiA | Resistance to competence killing |
| *Vibrio fischeri* | 3-Oxo-$C_6$-HSL | LuxI/LuxR | Bioluminescence |
| *Vibrio harveyi* | 3-Hydroxy-$C_4$-HSL | LuxLM/LuxN Lux-/LuxPQ | Bioluminescence |
| *Vibrio anguillarum* | 3-Oxo-$C_{10}$-HSL | VanI/VanR | – |
| *Yersinia enterocolitica* | $C_6$-HSL | YenI/YenR | – |
| *Yersinia pseudotuberculosis* | $C_8$-HSL | YtbI/YtbR | Bacterial aggregation, motility |

the LuxR-autoinducers complexes [4–6]. In general, this type of circuit is observed in different gram-negative bacteria with few exceptions (i.e. *M.xanthus*, *V. harveyi*) [2] (see more details in [1, 7–9]). We discuss some well understood quorum sensing circuits of gram-negative bacteria in this section.

### 2.2.1 Quorum Sensing Circuit of *Vibrio fischeri*

It has been observed that the *V. fischeri* has symbiotic relationship with the eukaryotic host. This bacterium lives in a nutrient rich environment and the cell density grows inside the light organ of the host [10–12]. In the signalling cascade, we observed two regulatory protein such as LuxI and LuxR. LuxI activates the production of *N*-(3-oxohexanoyl)- homoserine lactone (autoinducers of *V. fischeri*) and LuxR binds with *N*-(3-oxohexanoyl)- homoserine lactone. The interaction between LuxR and autoinducers exposes the LuxR DNA binding domain, which allows LuxR to combine with *luxICDABE* promoter and activate transcription of the *luxICDABE* operon [4, 13–17]. The LuxR-autoinducer complex behaves as a negative feedback loop (i.e. luxR expression), which decreases the positive feedback loop (i.e. *luxICDABE* expression) [4]. The concentration of autoinducers is same in intercellular as well as extracellular environment, because *N*-(3-oxohexanoyl)-homoserine lactone is easily diffusible across the cell membrane [18]. *V. fischeri* culture grows over the time and cell density reaches around $10^{11}$ cells/ml [19]. The autoinducers concentration reaches a threshold level (around 1–10 μg/ml) [20] and starts communication with other bacteria inside the host. So, the cell density is correlated with light production. Luciferase enzymes are needed for the production of light in these bacteria, which are encoded by *luxCDABE* (being as a part of

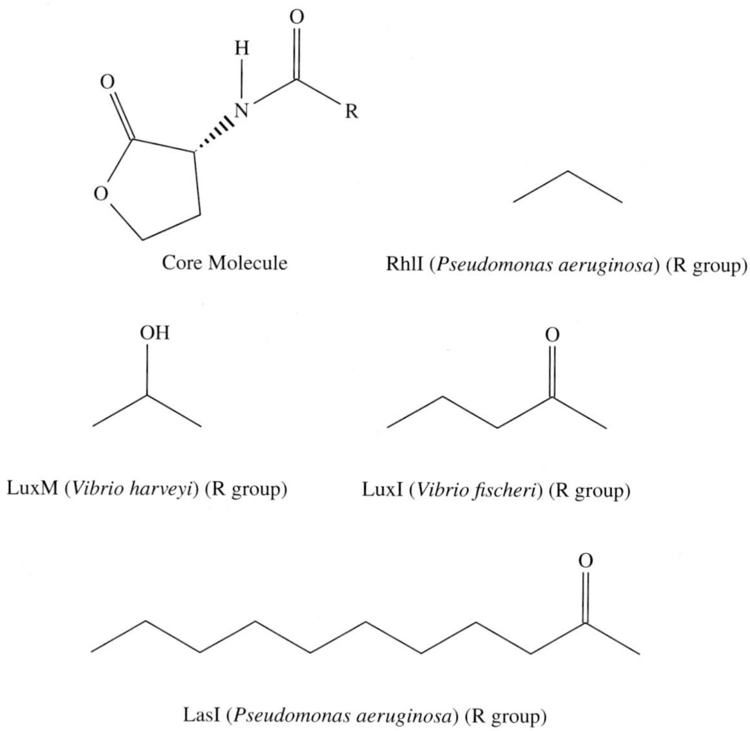


**Fig. 2.2** Chemical structures: The core molecule and R groups of some Acyl-homoserine lactones (autoinducers)

*luxICDABE* operon) [4, 21] (Fig. 2.3). This light production feature is known as bioluminescence. Eukaryotic host utilizes this light for particular purposes such as attracting preys and staying away from predators [22]. For example, *Monocentris japonicus* uses this *V. fischeri* light to attract a mate and *Euprymna scolopes* uses this same lightning feature of *V. fischeri* for antipredation strategy [2].

### 2.2.2 Quorum Sensing Circuit of *Pseudomonas aeruginosa*

*Pseudomonas aeruginosa* is a well known pathogenic bacteria, which has a hierarchical LuxI/R quorum sensing process. *P. aeruginosa* is responsible for the lung disease called cystic fibrosis and also regulate the biofilm formation [2]. Quorum sensing system of this bacteria has two signalling cascade such as LasI/LasR [23] and RhlI/RhlR [24] (both pairs are LuxI/LuxR homologues). LasI and RhlI produce autoinducers *N*-(3-oxododecanoyl)-homoserine lactone [25] and *N*-(butryl)-homoserine lactone [26], respectively, to regulate the quorum sensing circuit and control virulence genes. LasR binds with *N*-(3-oxododecanoyl)-homoserine lactone (autoinducer) and the complex (LasR-autoinducer) binds with

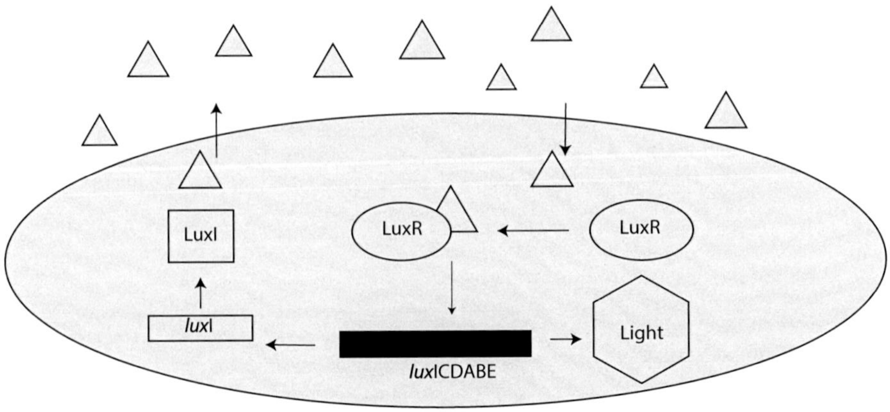


**Fig. 2.3** Illustration of quorum sensing circuit of *Vibrio fischeri* (LuxI/LuxR): The oval shape shows a bacterial cell. This system consists of two regulatory genes (*luxI* and *luxR*) and five luciferase structural genes (*luxCDABE*). The triangles are autoinducers. LuxI (protein) produces autoinducers. The concentration of autoinducers increases, when the cell population density rises. When the concentration of autoinducers reaches a certain level LuxR (protein) binds with autoinducers. LuxR-autoinducers complex binds with promoter region of *luxICDABE* and active the transcription process of the operon *luxICDABE* and produce light

the promoter region before the genes encoding virulence factors (i.e. alkaline phosphatase, exotoxinA, protease and elastase are encoded by *aprA*, *toxA*, *lasA* and *lasB*, respectively) [1, 23, 27, 28]. The infection mechanism of the host begins and is controlled by these secreted virulence factors. A positive feedback loop is observed, when the complex (LasR-autoinducer) triggers *lasI* expression [29].

In other signalling cascade, *rhlR* expression is activated by the complex (LasR-autoinducer). RhlI produces *N*-(butryl)-homoserine lactone (autoinducer) and RhlR binds with the autoinducer [30]. Two genes expressions (*lasB* and *aprA*) are also controlled by the complex (RhlR-autoinducer). Moreover, RhlR-autoinducer complex triggers specific genes such as *rpoS*, *rhlAB* and *lecA* [1, 8, 9, 24, 30–37]. We can observe an autoregulatory loop in the system (activation of *rhlI*). Both the signalling mechanisms (RhlI/RhlR and LasI/LasR) work sequentially (Fig. 2.4).

Beside this above mention signalling cascades, *P. aeruginosa* uses 2-heptyl-3-hydroxy-4-quinolone (also known as *Pseudomonas* quinolone signal (PQS). PQS is considered as an additional link between Rhl and Las circuits and partially controls *lasB* gene expression [38].

### 2.2.3 Quorum Sensing Mechanism of *Agrobacterium tumefaciens*

The crown gall tumours are induced by the plant pathogenic bacteria *Agrobacterium tumefaciens*. Bacterium transfers oncogenic Ti plasmid to the host for the formation

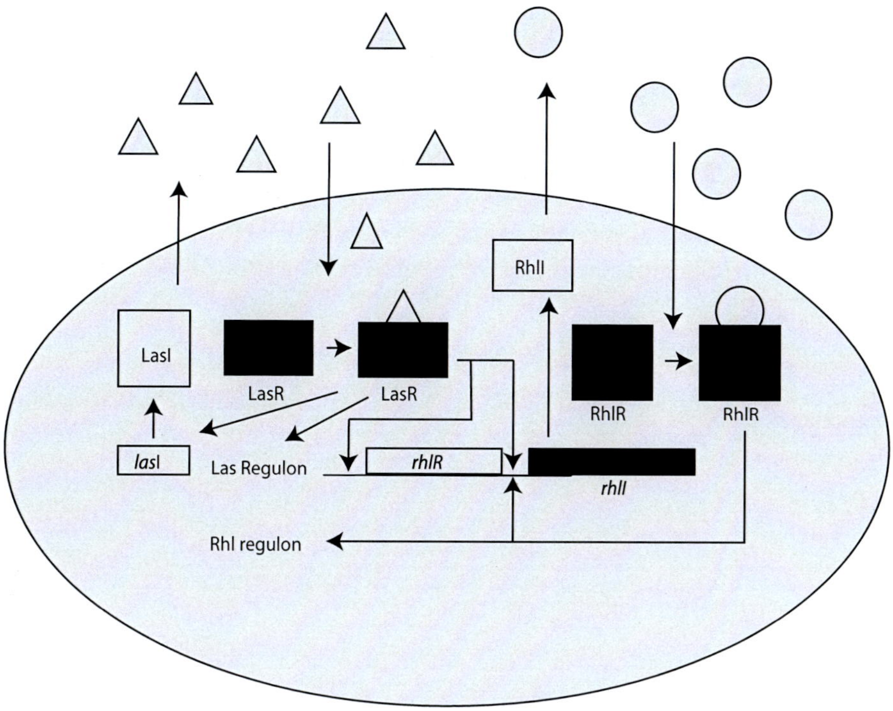


**Fig. 2.4** Quorum sensing circuit of *Pseudomonas aeruginosa*: The oval shape shows the bacterial cell. The triangle and the circle represent two different autoinducers such as *N*-(3-oxododecanoyl)-homoserine lactone and *N*-(butryl)-homoserine lactone, respectively. There are two signalling cascades (LasI/LasR and RhlI/RhlR). LasI produces *N*-(3-oxododecanoyl)-homoserine lactone (autoinducer) that binds to LasR. The complex (LasR-autoinducer) activates different targeted genes (including virulence genes), induces transcription of *rhlR* as well as initiates the second signalling cascade. RhlI also produces *N*-(butryl)-homoserine lactone (autoinducer) and RhlR binds with autoinducer. The RhlR-autoinducer complex triggers set of targeted genes

of tumour [39, 40]. Opines secretion in the plant and biosynthesis is controlled by the genes on the Ti plasmid. The conjugation between cells needs autoinducer signal and opine signal. Opines control the communication mechanism and are considered as nutrient source for bacteria. Opine regulates the TraR expression. Two different class of opine such as nopaline type and octapine type regulate conjugal Ti plasmids. *A. tumefaciens* quorum sensing circuit is very much similar with *V. fischeri* at low cell population density. Bacterium uses *N*-(3-oxoctanoyl)-homoserine lactone (autoinducer) for their communication [41,42]. We can observe TraI/TraR signalling cascade in this communication process. TraI produces autoinducers and TraR binds with autoinducers and forms a (TraR-autoinducer) complex, which induces the *traI* expression. In this way, a positive autoinduction loop is created. The complex (TraR-autoinducer) regulates *tra* operon, *trb* operon and *traM* gene [2, 43–45]. *trb* operon encodes necessary genes and *tra* operon triggers Ti plasmid mobilization.

Moreover, the complex (TraR-autoinducer) induces TraM and down regulates the communication process. TraM is an additional level of regulation in this quorum sensing circuit.

### 2.2.4 Quorum Sensing Mechanism of *Erwinia carotovora*

We can find soft rot in potato because of plant pathogenic bacteria *Erwinia carotovora* [46]. The quorum sensing process of *E. carotovora* consists of two signalling cascade ExpI/ExpR and CarI/CarR. ExpI/ExpR homologues to LuxI/LuxR that regulates the cascade to mount a victorious infection [2]. Exoenzymes secretion is controlled by ExpI/ExpR at high cell density. The second signalling cascade is CarI/CarR, which has a similarity with LuxI/R. ExpI and CarI both produce the same autoinducer known as *N*-(3-oxohexanoyl)-homoserine lactone [47]. ExpR and CarR response to the same biochemical signal. CarI/CarR system generates antibiotics as well [48, 49].

## 2.3 Quorum Sensing in Gram-Positive Bacteria

Gram-positive bacteria regulate the cell-to-cell communication process using oligopeptides (autoinducers). We observe a precursor protein in this system, which is translated from peptide signal precursor locus and divided into peptides (autoinducers). Peptides are transported via ABC transporter, because it is not diffusible across cell membrane. The autoinducers concentration increases and reaches the threshold concentration. Gram-positive bacteria have two-component histidine sensor kinases for detection of autoinducer. Then, we notice a series of phosphoryl events, which is initiated by peptide ligand. This phosphorylation triggers response regulator (DNA binding transcription process). Finally, targeted genes transcription is activated by the phosphorylated response regulator [2, 3, 7, 50–52]. Here, we are mainly discussing three gram-positive quorum sensing system (Figs. 2.5 and 2.6).

### 2.3.1 Quorum Sensing Process of *Streptococcus pneumoniae*

We observe genetic transformation in a gram-positive quorum sensing bacterium called *Streptococcus pneumoniae* [53]. This biochemical process needs that the bacterium becomes competent in order to get exogenous DNA molecules. This competent state is very complex phenomenon and partially controlled by cell-to-cell communication mechanism [54]. Competent state arises at the time of exponential growth. The *S. pneumoniae* loses the ability in later stage and departs from the competent state [53, 55, 56]. The competent state is developed by the signalling peptide known as competence stimulating peptide (CSP). ComC (41-amino acid precursor peptide) produces CPS (17-amino acid peptide) [57, 58]. This system

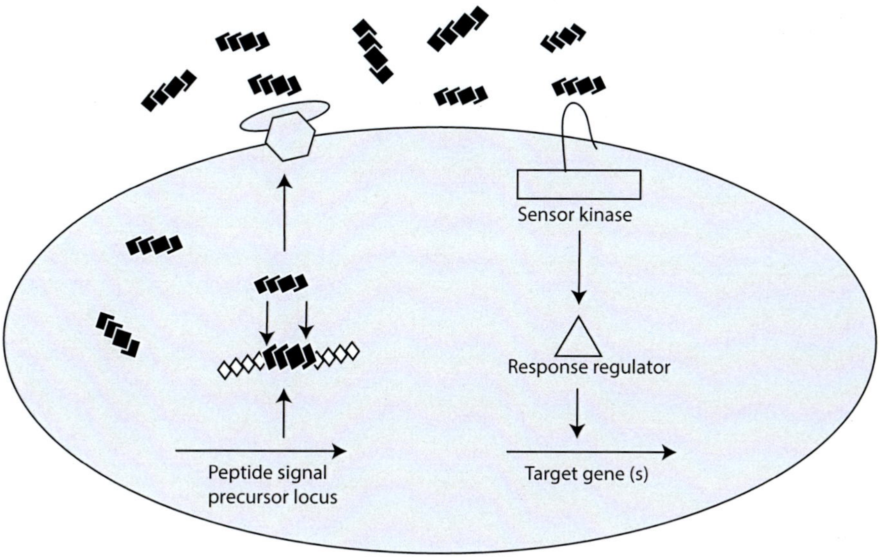


**Fig. 2.5** In general, schematic diagram of a quorum sensing system of a gram-positive bacteria. This quorum sensing mechanism is mediated by peptides. The oval shape represents bacterial cell. Black diamonds are signalling peptides (autoinducers). Precursor protein (black and white diamonds) is translated from a peptide signal precursor and generates autoinducers. These autoinducers transport through ABC transporter. Peptides (autoinducers) detected by sensor kinase, at high cell density and phosphoryl group is transferred to response regulator by autophosphorylation. The targeted genes are activated by phosphorylated response regulator

has ABC transporter, ComAB. ComAB secretes processed CSP [59, 60]. ComD is the sensor kinase protein, which can detect CSP at high cell density [61]. Autophosphorylation of ComD is induced by high level of CSP and phosphoryl group is transferred to ComE (response regulator). Finally, *comX* gene transcription is triggered by phospho-ComE [62].

### 2.3.2 Quorum Sensing Process of *Bacillus subtilis*

The peptide quorum sensing system is also observed in another gram-positive bacteria known as *Bacillus subtilis*. We notice competent state and sporulation mechanism, which are controlled by the two peptide mediated communication process. *B. subtilis* reaches the competent state at the transition between logarithmic and stationary phase growth [51,63]. When the bacteria live in limited nutrients condition and the environmental condition have also deteriorated, then the sporulation process occurs in *B. subtilis* [64]. Quorum sensing mechanism is mediated by two peptides, ComX and CSF (competence and sporulation factor). These peptides are ejected and the concentration of peptides (autoinducers) increases as the cell density rises. 55-amino acid precursor peptide generates ComX and ComQ is needed for

**Fig. 2.6** Chemical structures: Oligopeptide autoinducers

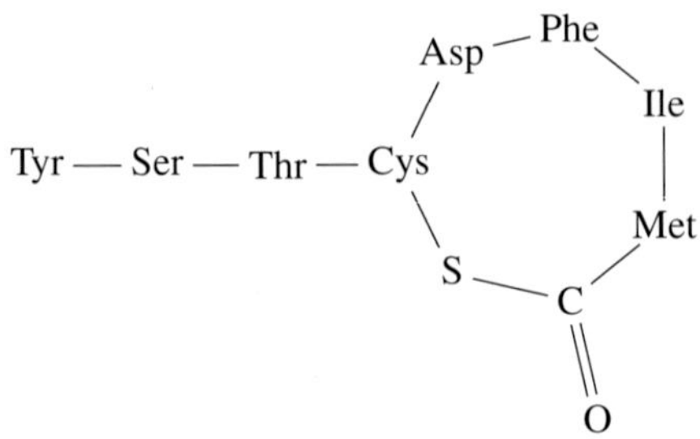


AIP-I (*Staphylococcus aureus*)

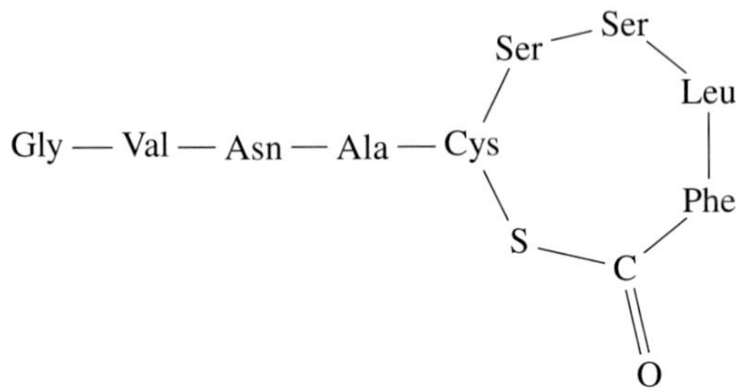


AIP-II (*Staphylococcus aureus*)

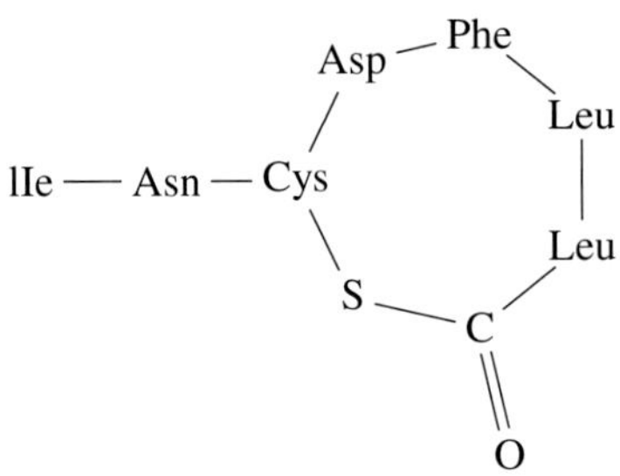


AIP-III (*Staphylococcus aureus*)

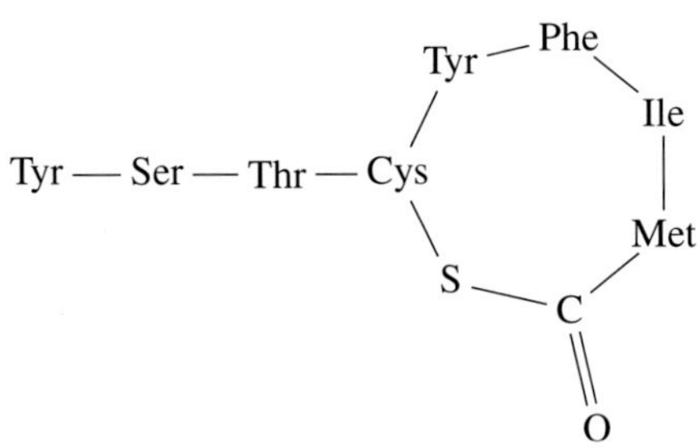


AIP-IV (*Staphylococcus aureus*)

production of ComX. ComP is a sensor kinase required for the detection of ComX. ComA is a response regulator of this signalling mechanism. The *comS* gene is activated by the phospho-ComA [65–68]. The degradation of ComK is inhibited by phospho-ComA. ComK is transcriptional activator associated with competence pathway.

*B. subtilis* also uses CFS (pentapeptide) to regulate the communication process. CSF is generated from the precursor peptide PhrC [66]. CSF is secreted via Opp (ABC type oligopeptide transporter). RapC (ComA-specific phosphatase) is inhibited by CSF (at low intracellular CSF concentration). *comS* gene expression is induced by CFS (at high intracellular CFS concentration) [66, 67, 69, 70]. So, competence is promoted at low intracellular CSF concentration, whereas sporulation is induced at high intracellular CSF concentration. RapB is inhibited by CSF, which dephosphorylates Spo0A (response regulator) and smooth the sporulation pathway [63, 70–72].

### 2.3.3 Quorum Sensing Mechanism of *Staphylococcus aureus*

*Staphylococcus aureus* is a gram-positive pathogenic bacteria. This is a multitalented bacterium, which causes several diseases such as endocarditis, toxic shock syndrome and skin infection. The *S. aureus* quorum sensing system is regulated by autoinducing peptide (AIP) [73]. We can also notice variation in AIPs. The density dependent pathogenicity is regulated by RNAIII (RNA molecule). RNAIII is partially controlled by *agrBDCA* operon. *agrBDCA* is transcribed from *hld* gene. *hld* encodes the RNAIII transcript. Octapeptide is produced from AgrD (precursor peptide). This production process depends on AgrB-dependent mechanism [74–80]. We observe a thio-lactone ring in AIP and a two competent system AgrC/ArgA (sensor kinase/ response regulator) which is this communication system [80–82]. The concentration of RNAIII is increased by phospho-AgrA. RNAIII triggers the gene expression as well as virulence factors.

## 2.4 Cross-Species Cell-to-Cell Communication

Bacteria can talk with other bacterial species, which is formally known as interspecies or cross-species communication process. This notion arose with the finding of autoinducers-2 (AI-2) in *Vibrio harveyi*. *luxS* gene is needed for AI-2 production and LuxS synthesis the AI-2. Bacteria use AI-2 based quorum sensing mechanism for interspecies cell-to-cell communication [7, 83, 84]. For example, *V. harveyi* lives in a mixed population (with other bacterium) and communicates with each other using two different type of autoinducers (AI-1 and AI-2). Bacteria use AI-1 for intraspecies communication and AI-2 for interspecies communication [83, 84]. There are several number of gram-negative and gram-positive bacteria that contain luxS gene (required for interspecies communication), such as *B. subtilis*, *S. aureus*, *E. coli*, *V. cholerae*, *Y. pestis*, *S. paratyphi*, *H. influenzae*, *K. pneumoniae*, *M.*

*tuberculosis* and many more [7, 84]. LuxS generates DPD (4,5-dihydroxy-2,3-pentonedione). DPD is highly reactive and derives signalling molecules AI-2 [3].

So, we conclude that bacteria can talk to each other (intraspecies and inter-species) using different types of chemical signalling molecules for their own survival strategies. Gram-negative bacteria use acyl-homoserine lactones (autoinducers) and gram-positive bacteria use peptide for regulating the quorum sensing systems. We will see how bacteria can regulate other biochemical phenomena such as biofilm formation, virulence, swarming and many more (with mathematical modelling approach) in the next couple of chapters.

## References

1. De Kievit TR, Iglewski BH (2000) Bacterial quorum sensing in pathogenic relationships. Infect Immu 68(9):4839–4849
2. Miller MB, Bassler BL (2001) Quorum sensing in bacteria. Annu Rev Microbiol 55(1):165–199
3. Waters CM, Bassler BL (2005) Quorum sensing: cell-to-cell communication in bacteria. Annu Rev Cell Dev Biol 21:319–346
4. Engebrecht J, Nealson K, Silverman M (1983) Bacterial bioluminescence: isolation and genetic analysis of functions from *Vibrio Fischeri*. Cell 32(3):773–781
5. Engebrecht J, Silverman M (1984) Identification of genes and gene products necessary for bacterial bioluminescence. Proc Nat Acad Sci 81(13):4154–4158
6. Engebrecht J, Silverman M (1987) Nucleotide sequence of the regulatory locus controlling expression of bacterial genes for bioluminescence. Nucleic Acids Res 15(24):10455–10467
7. Bassler BL (1999) How bacteria talk to each other: regulation of gene expression by quorum sensing. Curr Opin Microbiol 2(6):582–587
8. Fuqua C, Winans SC, Greenberg EP (1996) Census and consensus in bacterial ecosystems: the LuxR–LuxI family of quorum-sensing transcriptional regulators. Annu Rev Microbiol 50(1):727–751
9. Parsek MR, Greenberg EP (2000) Acyl-homoserine lactone quorum sensing in gram-negative bacteria: a signaling mechanism involved in associations with higher organisms. Proc Nat Acad Sci 97(16):8789–8793
10. Ruby EG (1996) Lessons from a cooperative, bacterial-animal association: the *Vibrio fischeri–Euprymna scolopes* light organ symbiosis. Annu Rev Microbiol 50(1):591–624
11. Ruby EG, McFall-Ngai MJ (1992) A squid that glows in the night: development of an animal-bacterial mutualism. J Bacteriol. 174(15):4865–4870
12. Visick KL, McFall-Ngai MJ (2000) An exclusive contract: specificity in the *Vibrio fischeri–Euprymna scolopes* partnership. J Bacteriol 182(7):1779–1787
13. Hanzelka BL, Greenberg EP (1995) Evidence that the N-terminal region of the *Vibrio fischeri* LuxR protein constitutes an autoinducer-binding domain. J Bacteriol 177(3):815–817
14. Schaefer AL, Hanzelka BL, Eberhard A, Greenberg EP (1996) Quorum sensing in *Vibrio fischeri*: probing autoinducer-LuxR interactions with autoinducer analogs. J Bacteriol 178(10):2897–2901
15. Stevens AM, Dolan KM, Greenberg EP (1994) Synergistic binding of the *Vibrio fischeri* LuxR transcriptional activator domain and RNA polymerase to the lux promoter region. Proc Nat Acad Sci 91(26):12619–12623
16. Stevens AM, Fujita N, Ishihama A, Greenberg EP (1999) Involvement of the RNA polymerase $\alpha$-subunit C-terminal domain in LuxR-dependent activation of the *Vibrio fischeri* luminescence genes. J Bacteriol 181(15):4704–4707

17. Stevens AM, Greenberg EP (1997) Quorum sensing in *Vibrio fischeri*: essential elements for activation of the luminescence genes. J Bacteriol 179(2):557–562
18. Kaplan HB, Greenberg EP (1985) Diffusion of autoinducer is involved in regulation of the *Vibrio fischeri* luminescence system. J Bacteriol 163(3):1210–1214
19. Nyholm SV, McFall-Ngai MJ (1998) Sampling the light-organ microenvironment of *Euprymna scolopes*: description of a population of host cells in association with the bacterial symbiont *Vibrio fischeri*. Biol Bull 195(2):89–97
20. Eberhard A, Burlingame AL, Eberhard C, Kenyon GL, Nealson KH, Oppenheimer NJ (1981) Structural identification of autoinducer of *Photobacterium fischeri* luciferase. Biochemistry 20(9):2444–2449
21. Lee CY, Szittner RB, Miyamoto CM, Meighen EA (1993) The gene convergent to luxG in *Vibrio fischeri* codes for a protein related in sequence to RibG and deoxycytidylate deaminase. Biochim Biophys Acta Biomembr 1143(3):337–339
22. Nealson KH, Hastings JW (1979) Bacterial bioluminescence: its control and ecological significance. Microbiol Rev 43(4):496
23. Passador L, Cook JM, Gambello MJ, Rust L, Iglewski BH (1993) Expression of *Pseudomonas aeruginosa* virulence genes requires cell-to-cell communication. Science 260(5111):1127–1130.
24. Brint JM, Ohman DE (1995) Synthesis of multiple exoproducts in *Pseudomonas aeruginosa* is under the control of RhlR-RhlI, another set of regulators in strain PAO1 with homology to the autoinducer-responsive LuxR–LuxI family. J Bacteriol 177(24):7155–7163
25. Pearson JP, Gray KM, Passador L, Tucker KD, Eberhard A, Iglewski BH, Greenberg EP (1994) Structure of the autoinducer required for expression of *Pseudomonas aeruginosa* virulence genes. Proc Nat Acad Sci 91(1):197–201
26. Pearson JP, Passador L, Iglewski BH, Greenberg EP (1995) A second N-acylhomoserine lactone signal produced by *Pseudomonas aeruginosa*. Proc Nat Acad Sci 92(5):1490–1494
27. Davies DG, Parsek MR, Pearson JP, Iglewski BH, Costerton JW, Greenberg EP (1998) The involvement of cell-to-cell signals in the development of a bacterial biofilm. Science 280(5361):295–298
28. Jones S, Yu B, Bainton NA, Birdsall M, Bycroft BW, Chhabra SR, Cox AJ, Golby P, Reeves PJ, Stephens S (1993) The lux autoinducer regulates the production of exoenzyme virulence determinants in *Erwinia carotovora* and *Pseudomonas aeruginosa*. EMBO J 12(6):2477–2482
29. Seed PC, Passador L, Iglewski BH (1995) Activation of the *Pseudomonas aeruginosa* lasI gene by LasR and the *Pseudomonas autoinducer* PAI: an autoinduction regulatory hierarchy. J Bacteriol 177(3):654–659
30. Pesci EC, Pearson JP, Seed PC, Iglewski BH (1997) Regulation of *las* and *rhl* quorum sensing in *Pseudomonas aeruginosa*. J Bacteriol 179(10):3127–3132
31. Hassett DJ, Ma JF, Elkins JG, McDermott TR, Ochsner UA, West SE, Huang CT, Fredericks J, Burnett S, Stewart PS, McFeters G (1999) Quorum sensing in *Pseudomonas aeruginosa* controls expression of catalase and superoxide dismutase genes and mediates biofilm susceptibility to hydrogen peroxide. Mol Microbiol 34(5):1082–1093
32. Latifi A, Foglino M, Tanaka K, Williams P, Lazdunski A (1996) A hierarchical quorum-sensing cascade in *Pseudomonas aeruginosa* links the transcriptional activators LasR and RhIR (VsmR) to expression of the stationary-phase sigma factor RpoS. Mol Microbiol 21(6):1137–1146
33. Parsek MR, Greenberg EP (1999) [3] Quorum sensing signals in development of *Pseudomonas aeruginosa* biofilms. In: Methods in enzymology, vol 310. Academic Press, pp 43–55
34. Pearson JP, Pesci EC, Iglewski BH (1997) Roles of *Pseudomonas aeruginosa las* and *rhl* quorum-sensing systems in control of elastase and rhamnolipid biosynthesis genes. J Bacteriol 179(18):5756–5767
35. Whiteley M, Lee KM, Greenberg EP (1999) Identification of genes controlled by quorum sensing in *Pseudomonas aeruginosa*. Proc Nat Acad Sci 96(24):13904–13909
36. Whiteley M, Parsek MR, Greenberg EP (2000) Regulation of quorum sensing by RpoS in *Pseudomonas aeruginosa*. J Bacteriol 182(15):4356–4360

37. Winzer K, Falconer C, Garber NC, Diggle SP, Camara M, Williams P (2000) The *Pseudomonas aeruginosa* lectins PA-IL and PA-IIL are controlled by quorum sensing and by RpoS. J Bacteriol 182(22):6401–6411
38. Pesci EC, Milbank JB, Pearson JP, McKnight S, Kende AS, Greenberg EP, Iglewski BH (1999) Quinolone signaling in the cell-to-cell communication system of *Pseudomonas aeruginosa.* Proc Nat Acad Sci 96(20):11229–11234
39. Christie PJ (1997) *Agrobacterium tumefaciens* T-complex transport apparatus: a paradigm for a new family of multifunctional transporters in eubacteria. J Bacteriol 179(10):3085–3094
40. Sheng J, Citovsky V (1996) Agrobacterium-plant cell DNA transport: have virulence proteins, will travel. Plant Cell 8(10):1699–1710
41. Dessaux Y, Petit A, Tempé J (1992) Opines in Agrobacterium biology. Molecular signals in plant-microbe communications. CRC Press, Boca Raton, pp 109–136
42. Zhang L, Murphy PJ, Kerr A, Tate ME (1993) Agrobacterium conjugation and gene regulation by N-acyl-L-homoserine lactones. Nature 362(6419):446–448
43. Fuqua C, Burbea M, Winans SC (1995) Activity of the Agrobacterium Ti plasmid conjugal transfer regulator TraR is inhibited by the product of the traM gene. J Bacteriol 177(5):1367–1373
44. Hwang I, Li PL, Zhang L, Piper KR, Cook DM, Tate ME, Farrand SK (1994) TraI, a LuxI homologue, is responsible for production of conjugation factor, the Ti plasmid N-acylhomoserine lactone autoinducer. Proc Nat Acad Sci 91(11):4639–4643
45. Piper KR, von Bodman SB, Farrand SK (1993) Conjugation factor of *Agrobacterium tumefaciens* regulates Ti plasmid transfer by autoinduction. Nature 362(6419):448
46. Hinton JC, Sidebotham JM, Hyman LJ, Pérombelon MC, Salmond GP (1989) Isolation and characterisation of transposon-induced mutants of *Erwinia carotovora* subsp. *atroseptica* exhibiting reduced virulence. Mol Gen Genet 217(1):141–148
47. Andersson RA, Eriksson AR, Heikinheimo R, Mäe A, Pirhonen M, Kõiv V, Hyytiäinen H, Tuikkala A, Palva ET (2000) Quorum sensing in the plant pathogen *Erwinia carotovora* subsp. *carotovora*: the role of expREcc. Mol Plant-Microbe Interact 13(4):384–393
48. Bainton NJ, Stead P, Chhabra SR, Bycroft BW, Salmond GPC, Stewart GS, Williams P (1992) N-(3-oxohexanoyl)-L-homoserine lactone regulates carbapenem antibiotic production in *Erwinia carotovora.* Biochem J 288(3):997–1004
49. Williams P, Bainton NJ, Swift S, Chhabra SR, Winson MK, Stewart GS, Salmond GP, Bycroft BW (1992) Small molecule-mediated density-dependent control of gene expression in prokaryotes: bioluminescence and the biosynthesis of carbapenem antibiotics. FEMS Microbiol Lett 100(1–3):161–167
50. Kleerebezem M, Quadri LE, Kuipers OP, De Vos WM (1997) Quorum sensing by peptide pheromones and two-component signal-transduction systems in Gram-positive bacteria. Mol Microbiol 24(5):895–904
51. Lazazzera BA, Grossman AD (1998) The ins and outs of peptide signaling. Trends Microbiol 6(7):288–294
52. Bourret RB (1995) Two-component signal transduction, vol 2. In: Hoch JA, Silhavy TJ (eds). ASM Press, Washington
53. Dawson MH, Sia RH (1931) In vitro transformation of pneumococcal types: I. A technique for inducing transformation of pneumococcal types in vitro. J Exp Med 54(5):681
54. Håvarstein LS, Morrison DA (1999) Quorum sensing and peptide pheromones in streptococcal competence for genetic transformation. In: Cell–cell signaling in bacteria. ASM Press, Washington, pp 9–26
55. Hotchkiss RD (1954) Cyclical behavior in pneumococcal growth and transformability occasioned by environmental changes. Proc Nat Acad Sci USA 40(2):49–55
56. Tomasz A, Hotchkiss RD (1964) Regulation of the transformability of pneumococcal cultures by macromolecular cell products. Proc Nat Acad Sci USA 51(3):480–487
57. Håvarstein LS, Coomaraswamy G, Morrison DA (1995) An unmodified heptadecapeptide pheromone induces competence for genetic transformation in *Streptococcus pneumoniae.* Proc Nat Acad Sci 92(24):11140–11144

58. Pozzi G, Masala L, Iannelli F, Manganelli R, Håvarstein LS, Piccoli L, Simon D, Morrison DA (1996) Competence for genetic transformation in encapsulated strains of *Streptococcus pneumoniae*: two allelic variants of the peptide pheromone. J Bacteriol 178(20):6087–6090
59. Hui FM, Morrison DA (1991) Genetic transformation in *Streptococcus pneumoniae*: nucleotide sequence analysis shows ComA, a gene required for competence induction, to be a member of the bacterial ATP-dependent transport protein family. J Bacteriol 173(1):372–381
60. Hui FM, Zhou L, Morrison DA (1995) Competence for genetic transformation in *Streptococcus pneumoniae*: organization of a regulatory locus with homology to two lactococcin A secretion genes. Gene 153(1):25–31
61. Pestova EV, Håvarstein LS, Morrison DA (1996) Regulation of competence for genetic transformation in *Streptococcus pneumoniae* by an auto-induced peptide pheromone and a two-component regulatory system. Mol Microbiol 21(4):853–862
62. Lee MS, Morrison DA (1999) Identification of a new regulator in *Streptococcus pneumoniae* linking quorum sensing to competence for genetic transformation. J Bacteriol 181(16):5004–5016
63. Grossman AD (1995) Genetic networks controlling the initiation of sporulation and the development of genetic competence in *Bacillus subtilis*. Annu Rev Genet 29(1):477–508
64. Hoch JA (1995) Control of cellular development in sporulating bacteria by the phosphorelay two-component signal transduction system. In: Two-component signal transduction. American Society of Microbiology, Washington, pp. 129–144
65. Magnuson R, Solomon J, Grossman AD (1994) Biochemical and genetic characterization of a competence pheromone from *B. subtilis*. Cell 77(2):207–216
66. Solomon JM, Lazazzera BA, Grossman AD (1996) Purification and characterization of an extracellular peptide factor that affects two different developmental pathways in *Bacillus subtilis*. Genes Dev 10(16):2014–2024
67. Solomon JM, Magnuson R, Srivastava A, Grossman AD (1995) Convergent sensing pathways mediate response to two extracellular competence factors in *Bacillus subtilis*. Genes Dev 9(5):547–558
68. Turgay K, Hahn J, Burghoorn J, Dubnau D (1998) Competence in *Bacillus subtilis* is controlled by regulated proteolysis of a transcription factor. EMBO J 17(22), 6730–6738.
69. Lazazzera BA, Solomon JM, Grossman AD (1997) An exported peptide functions intracellularly to contribute to cell density signaling in *B. subtilis*. Cell **89**(6), 917–925
70. Perego M (1997) A peptide export–import control circuit modulating bacterial development regulates protein phosphatases of the phosphorelay. Proc Nat Acad Sci 94(16):8612–8617
71. Hoch JA (1993) Regulation of the phosphorelay and the initiation of sporulation in *Bacillus subtilis*. Annu Rev Microbiol 47(1):441–465
72. Perego M, Hanstein C, Welsh KM, Djavakhishvili T, Glaser P, Hoch JA (1994) Multiple protein-aspartate phosphatases provide a mechanism for the integration of diverse signals in the control of development in *B. subtilis*. Cell 79(6):1047–1055
73. Novick RP (1999) Regulation of pathogenicity in *Staphylococcus aureus* by a peptide-based density-sensing system. In: Cell-cell signaling in bacteria. American Society for Microbiology, Washington, pp 129–146
74. Janzon L, Arvidson S (1990) The role of the delta-lysin gene (hld) in the regulation of virulence genes by the accessory gene regulator (agr) in *Staphylococcus aureus*. EMBO J **9**(5):1391–1399
75. Ji G, Beavis R, Novick RP (1997) Bacterial interference caused by autoinducing peptide variants. Science 276(5321):2027–2030
76. Ji G, Beavis RC, Novick RP (1995) Cell density control of staphylococcal virulence mediated by an octapeptide pheromone. Proc Nat Acad Sci 92(26):12055–12059
77. Morfeldt E, Janzou L, Arvidson S, Löfdahl S (1988) Cloning of a chromosomal locus (exp) which regulates the expression of several exoprotein genes in *Staphylococcus aureus*. Mol Gen Genet 211(3):435–440

78. Morfeldt E, Taylor DV, Von Gabain A, Arvidson S (1995) Activation of alpha-toxin translation in *Staphylococcus aureus* by the trans-encoded antisense RNA, RNAIII. EMBO J 14(18):4569–4577.
79. Novick RP, Projan SJ, Kornblum J, Ross HF, Ji G, Kreiswirth B, Vandenesch F, Moghazeh S (1995) Theagr P2 operon: an autocatalytic sensory transduction system in *Staphylococcus aureus*. Mol Gen Genet 248(4):446–458
80. Peng HL, Novick RP, Kreiswirth B, Kornblum J, Schlievert PM (1988) Cloning, characterization, and sequencing of an accessory gene regulator (agr) in *Staphylococcus aureus*. J Bacteriol 170(9):4365–4372
81. Lina G, Jarraud S, Ji G, Greenland T, Pedraza A, Etienne J, Novick RP, Vandenesch F (1998) Transmembrane topology and histidine protein kinase activity of AgrC, the agr signal receptor in *Staphylococcus aureus*. Mol Microbiol 28(3):655–662
82. Mayville P, Ji G, Beavis R, Yang H, Goger M, Novick RP, Muir TW (1999) Structure-activity analysis of synthetic autoinducing thiolactone peptides from *Staphylococcus aureus* responsible for virulence. Proc Nat Acad Sci 96(4):1218–1223
83. Bassler BL, Greenberg EP, Stevens AM (1997) Cross-species induction of luminescence in the quorum-sensing bacterium *Vibrio harveyi*. J Bacteriol 179(12):4043–4045
84. Surette MG, Miller MB, Bassler BL (1999) Quorum sensing in *Escherichia coli*, *Salmonella typhimurium*, and *Vibrio harveyi*: a new family of genes responsible for autoinducer production. Proc Nat Acad Sci 96(4):1639–1644

# Mathematical Models of Quorum Sensing Molecular Mechanisms

# 3

**Abstract**

Bacterial cell communication process is modelled mathematically with emphasis on the molecular mechanism of the gram-negative and gram-positive bacteria. This chapter mainly focuses on the mathematical formalism of the *lux* regulatory model, up-down regulatory model, Dockery-Keener model, complex formation model and the dynamical model of *Staphylococcus aureus* cell communication. All these type of mathematical modelling approaches give us a significant and fundamental insight of the bacterial communication process at the molecular level.

## 3.1 *Lux* Regulation Model

We observe LuxI/R types regulatory process in various quorum sensing bacteria. Bacteria use acyl-homoserine lactones (autoinducers) to regulate the cell-to-cell communication systems. *V. fischeri* is one of them, who use LuxI/R type regulatory process and produce light (See details in Chap. 2). The lux regulon of *V. fischeri* consists of two parts such as right operon and left operon. The right operon contains *luxI* and left operon contains *luxR* as well. *luxR* gene encodes a transcriptional activator. The *lux* box lies between two operons and activates both operons transcription. LuxR binds with autoinducers and forms a complex. James et al. [1] proposed a fist mathematical model of the *lux* regulatory system in *V. fischeri* and analysed the *lux* gene regulatory mechanism.

### 3.1.1 Model Descriptions

Now we introduce a mathematical model of the lux regulatory system in *V. fischeri*. The analysis of the system mainly focuses on the lux gene regulation (in a single


S. Majumdar, S. Roy, *Microbial Communication*,
https://doi.org/10.1007/978-981-15-7417-7_3

cell) and compares the situations with and without autoinducers in the external environment. This mathematical framework has some basic assumptions as follows [1],

- Autoinducer $A$ binds with LuxR $R$ and forms a complex $C$. This reaction consists of $k_1$ (binding rate constant) and $k_2$ (dissociation rate constant), which follows that $k_1\hat{A}\hat{R}$ is binding reaction rate and $k_2\hat{C}$ is dissociation reaction rate, where $\hat{A}$, $\hat{R}$ and $\hat{C}$ represent the concentration of $A$, $R$ and $C$.
- $\frac{f\hat{C}}{1+f\hat{C}}$ is the proportion of time, when the *lux* box is occupied by the complex (linear dependency on $\hat{C}$, say $f\hat{C}$).
- $q \times \frac{f\hat{C}}{1+f\hat{C}}$ and $p \times \frac{f\hat{C}}{1+f\hat{C}}$ are the LuxR and autoinducer synthesis rate, respectively ($p$ and $q$ are constants).
- $n\hat{A}$ and $b\hat{R}$ are the diffusion rate of autoinducer and degradation rate of LuxR, respectively ($b$ is a constant and $n$ is diffusion constant).

The mathematical model is based on the above assumptions and a single bacterium can emit autoinducer through the cell membrane. So, we can formulate a dynamical system, which shows the lux regulatory system as follows [1]

$$\frac{d\hat{A}}{dt} = k_2\hat{C} - k_1\hat{A}\hat{R} - n\hat{A} + p\frac{f\hat{C}}{1+f\hat{C}} \tag{3.1}$$

$$\frac{d\hat{R}}{dt} = k_2\hat{C} - k_1\hat{A}\hat{R} - b\hat{R} + q\frac{f\hat{C}}{1+f\hat{C}} \tag{3.2}$$

$$\frac{d\hat{C}}{dt} = k_1\hat{A}\hat{R} - k_2\hat{C} \tag{3.3}$$

There are three coupled differential equations, which represent the rate of concentration of autoinducer ($\frac{d\hat{A}}{dt}$), LuxR ($\frac{d\hat{R}}{dt}$) and complex ($\frac{d\hat{C}}{dt}$). We compare with and without autoinducer in the external environment by modifying the Eq. 3.1 as below

$$\frac{d\hat{A}}{dt} = k_2\hat{C} - k_1\hat{A}\hat{R} - n\left(\hat{A} - \hat{A}_{ex}\right) + p\frac{f\hat{C}}{1+f\hat{C}} \tag{3.4}$$

where $\hat{A}_{ex}$ is the extracellular autoinducers concentration.

### 3.1.2 Discussion of *lux* Regulation Model

The dynamical system (Eqs. 3.1–3.2) shows the lux circuit (without autoinducers in the extracellular environment). We find three steady state in the system, among them two are stable and one is unstable. The first steady state (say $S_1$) is a non-induced state of luminescence. To sustain a stable equilibrium, the complex has to reach a threshold concentration. Let us say, other steady states are $S_2$ (unstable) and $S_3$ (stable) (check $\hat{C} > \frac{1}{f}$ for stability). The quorum sensing system (regulated by LuxI/R) can reach the stable steady state $S_3$ with higher bioluminescence. $S_2$ is unsustainable, because of instability. So, we notice a switching behaviour in the *lux* system. In the presence of autoinducers in the extracellular environment, the stability of the system is not affected greatly. Moreover, the model speculates that bioluminescence is controlled by gene expression under certain conditions (without presence of autoinducers in the extracellular environment) [1].

## 3.2 Up-down Regulatory Model

Quorum sensing is also considered as a density dependent widespread bacterial collective behaviour controlled by autoinducers (also known as quorum sensing molecule (QSM)). This bacterial population can be viewed as two subpopulation such as down-regulated and up-regulated. Ward et al. [2] consider the regulatory mechanism of the bacteria (*V. fischeri*) in a well mixed system and study the bacterial growth using mathematical formalism. Moreover, this mathematical framework gives some essential measurable macroscopic features on cell-to-cell communication process.

We consider the *V. fischeri* communication process, where QSM binds with the appropriate protein to form a complex. Then, complex binds to the lux-box part of quorum sensing gene region of the chromosome. The binding of the lux-box induces activation of quorum sensing genes from a down-regulated state to up-regulated one[2].

### 3.2.1 Mathematical Model

Mathematical model is based on the biologically compatible assumptions given below,

- The bacterial population is made up of down-regulated (with cell number density $\hat{N}_d$) and up-regulated (with cell number density $\hat{N}_u$) subpopulation of cells, corresponding to empty lux-box or complex bound, respectively.
- $\hat{A}$ is the autoinducer concentration.
- Down-regulated cells are up-regulated by autoinducers with constant rate $\hat{\alpha}$.
- Down-regulated and up-regulated cells produce autoinducers, at rates $\hat{k}_d$ and $\hat{k}_u$, respectively ($\hat{k}_d << \hat{k}_u$).
- Cell division rate is same for down-regulation and up-regulation, which is determined by the parameter $\hat{r}$.
- Two down-regulated cells are generated by a cell division of one down-regulated cell.

- Cell division of one up-regulated cell generates $\hat{\gamma}$ up-regulated and $(2-\hat{\gamma})$ down-regulated cell on average $(0 \leq \hat{\gamma} \leq 2)$. We expect that $(\hat{\gamma} \simeq 1)$, which implies that one up-regulated cell generates one up-regulated and one down-regulated cell as well.
- Down-regulated occurs spontaneously, due to breakdown of lux-box bound QSM-QSP complex at a rate $\hat{\beta}$.
- Autoinducers can be broken down by the medium, and hence lost to the system, at a rate $\hat{\lambda}$.

We get a nonlinear dynamical system, based on the above assumptions,

$$\frac{d\hat{N}_d}{dt} = \hat{r}\left(\hat{N}_d + (2-\hat{\gamma})\hat{N}_u\right)\hat{F}\left(\hat{N}_d + \hat{N}_u\right) - \hat{\alpha}\hat{G}\left(\hat{A}\right)\hat{N}_d + \hat{\beta}\hat{N}_u \tag{3.5}$$

$$\frac{d\hat{N}_u}{dt} = \hat{r}\left(\hat{\gamma}-1\right)\hat{N}_u\hat{F}\left(\hat{N}_d + \hat{N}_u\right) + \hat{\alpha}\hat{G}\left(\hat{A}\right)\hat{N}_d - \hat{\beta}\hat{N}_u \tag{3.6}$$

$$\frac{d\hat{A}}{dt} = \hat{k}_u\hat{N}_u + \hat{k}_d\hat{N}_d - \hat{\alpha}\hat{G}\left(\hat{A}\right)\hat{N}_d - \hat{\lambda}\hat{A} \tag{3.7}$$

where $\hat{F}(.)$ represents a bacterial growth function $(\hat{F}(0) = 0)$ and $\hat{G}(\hat{A})$ shows overall complex formation and lux-box binding $(\hat{G}(\hat{A}) = \hat{A})$. The total bacterial cell number density is $\hat{N}_T = \hat{N}_d + \hat{N}_u$. So, we can simplify the dynamical system by adding equation (3.5) and (3.6). We have

$$\frac{d\hat{N}_T}{dt} = \hat{r}\hat{N}_T\hat{F}\left(\hat{N}_T\right) \tag{3.8}$$

Now, we assume $\hat{F}(\hat{N}_T = 1 - \frac{\hat{N}_T}{\hat{K}}$ (logistic growth), where $\hat{K}$ is the carrying capacity. Moreover, we consider $\hat{F}(\hat{N}_T$ is continuous. Then, we find a stable steady state at $\hat{N}_T = \hat{K}$ $(\hat{F}^{'} > 0$ and $\hat{F}^{'} < 0)$ and an unstable steady state at $\hat{N}_T = 0$.

### 3.2.2 Comments on Up-down Regulatory Model

The up-down regulatory model describes the bacterial quorum sensing mechanism. The model explained the production and activity of a single quorum sensing molecule with consequences. Model assumption is quite effective to describe the production of autoinducer and bacterial growth in batch culture. The mathematical formalism predicts switching behaviour in communication system which is similar with experiment. We can move onto several analysis with this up-down regulatory model which includes steady state analysis, linear stability analysis and asymptotic analysis to get more detail information about the system (see detail in [2]). It has been observed that general and stable solution of the model is compatible with real solution. One thing is not clear from the analysis of the model that what happens after quorum is reached.

This simple model considers two different states and single QSM. But biological systems are not simple as that. We consider this as a basic model and groom us towards more complex one with significant parameters. This kind of study gives us valuable deep insight into bacterial communication systems which can be implemented for the further research works in medical sciences.

## 3.3 Dockery-Keener Model

Jack Dockery and James Keener proposed a mathematical formalism of the bacterial communication process in *Pseudomonas aeruginosa* [3]. This model focuses on two biochemical regulatory systems such as *las* and *rhl* system (see Fig. 3.1). We call this mathematical model as Dockery-Keener model. Let us consider, variables $\hat{R}$, $\hat{A}$, $\hat{P}$, $\hat{L}$, $\hat{S}$, $\hat{r}$, $\hat{l}$, $\hat{s}$ as a concentrations of LasR, 3-oxo-C12-HSL, LasR/3-oxo-C12-HSL, LasI, RsaL, lasR mRNA, lasI mRNA, rsaL mRNA. Now, we are trying to find out kinetics of the system. We consider $\hat{k}_{\hat{P}}$ and $\hat{k}_{\hat{R}\hat{A}}$ as a degraded rate and the law of mass action rate of dimer $\hat{P}$,

$$\frac{d\hat{P}}{dt} = \hat{k}_{\hat{R}\hat{A}}\hat{R}\hat{A} - \hat{k}_{\hat{P}}\hat{P}$$

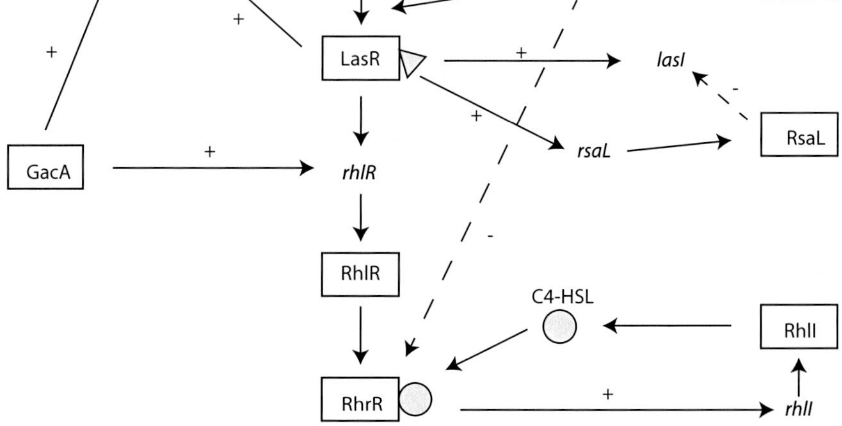


**Fig. 3.1** Schematic diagram of the Dockery-Keener model

Enzyme LasR and autoinducer are involved in the production of $\hat{P}$. Let us consider $\hat{k}_{\hat{R}}$ and $\hat{k}_{\hat{A}}$ be the natural degradation rate of enzyme LasR and autoinducer, respectively. Enzyme LasR and autoinducer are produced by the degradation of $\hat{P}$ and by lasR mRNA and lasI enzyme at rate $\hat{k}_1$ and $\hat{k}_2$, respectively,

$$\frac{d\hat{R}}{dt} = -\hat{k}_{\hat{R}\hat{A}}\hat{R}\hat{A} + \hat{k}_{\hat{P}}\hat{P} - \hat{k}_{\hat{R}}\hat{R} + \hat{k}_1\hat{r}$$

$$\frac{d\hat{A}}{dt} = -\hat{k}_{\hat{R}\hat{A}}\hat{R}\hat{A} + \hat{k}_{\hat{P}}\hat{P} + \hat{k}_2\hat{L} - \hat{k}_{\hat{A}}\hat{A}$$

LasI (enzyme)) is generated by lasI mRNA at rate $\hat{k}_3$ with a degrades rate $\hat{k}_{\hat{L}}$

$$\frac{d\hat{L}}{dt} = \hat{k}_3\hat{l} - \hat{k}_{\hat{L}}\hat{L}$$

RsaL (inhibitor) is generated by rsaL mRNA at rate $\hat{k}_4$ with degrades rate $\hat{k}_{\hat{S}}$

$$\frac{d\hat{S}}{dt} = \hat{k}_4\hat{s} - \hat{k}_{\hat{S}}\hat{S}$$

DNA produces all messenger RNA at rates of Michaelis–Menten type. So, rsaL mRNA, lasR mRNA, lasI mRNA are produced as follows,

$$\frac{d\hat{s}}{dt} = \hat{V}_{\hat{s}}\frac{\hat{P}}{\hat{K}_{\hat{s}} + \hat{P}} - \hat{k}_{\hat{s}}\hat{s} \;\left(\text{where } \hat{k}_{\hat{s}} \text{ is natural rate}\right)$$

$$\frac{d\hat{r}}{dt} = \hat{V}_{\hat{r}}\frac{\hat{P}}{\hat{K}_{\hat{r}} + \hat{P}} - \hat{k}_{\hat{r}}\hat{r} + \hat{r}_0 \;\left(\text{where } \hat{k}_{\hat{r}} \text{ is natural rate, } \hat{r}_0 \text{ is basal rate}\right)$$

$$\frac{d\hat{l}}{dt} = \hat{V}_{\hat{l}}\frac{\hat{P}}{\hat{K}_{\hat{l}} + \hat{P}}\frac{1}{\hat{K}_{\hat{S}} + \hat{S}} - \hat{k}_{\hat{l}}\hat{l} + \hat{l}_0 \;\left(\text{where } \hat{k}_{\hat{l}} \text{ is natural rate, } \hat{l}_0 \text{ is basal rate}\right)$$

Now, we simplify the system because some reactions are faster than others. $\hat{k}_{\hat{r}}$ and $\hat{k}_{\hat{l}}$ much larger than $\hat{k}_{\hat{L}}$ and $\hat{k}_{\hat{R}}$, respectively, because LasR mRNA and lasI mRNA have shorter lived than LasR and LasL. We get, quasi steady state $\hat{l}$ and $\hat{r}$,

$$\hat{k}_{\hat{l}}\hat{l} = \hat{V}_{\hat{l}}\frac{\hat{P}}{\hat{K}_{\hat{l}} + \hat{P}}\frac{1}{\hat{K}_{\hat{S}} + \hat{S}} + \hat{l}_0$$

$$\hat{k}_{\hat{r}}\hat{r} = \hat{V}_{\hat{r}}\frac{\hat{P}}{\hat{K}_{\hat{r}} + \hat{P}} + \hat{r}_0$$

Let $\hat{L}$ be a first order linear filter, which tracks $\hat{l}$ (with delay). We can ignore the delay as follow,

$$\hat{k}_3\hat{l} = \hat{k}_{\hat{L}}\hat{L}$$

The production of $\hat{l}$ is inhibited by $\hat{S}$ (not effective on quorum sensing). So, we can ignore this variable.

### 3.3.1 Mathematical Model (ODE)

We are now in the position to model the quorum sensing phenomena (molecular mechanisms) of *P. aeruginosa*, based on the above assumptions and mathematical frameworks. The governing set of equations is as follows [3],

$$\frac{d\hat{P}}{dt} = \hat{k}_{\hat{R}\hat{A}}\hat{R}\hat{A} - \hat{k}_{\hat{P}}\hat{P} \tag{3.9}$$

$$\frac{d\hat{R}}{dt} = -\hat{k}_{\hat{R}\hat{A}}\hat{R}\hat{A} + \hat{k}_{\hat{P}}\hat{P} - \hat{k}_{\hat{R}}\hat{R} + \hat{V}_{\hat{R}}\frac{\hat{P}}{\hat{K}_{\hat{R}} + \hat{P}} + \hat{R}_0 \tag{3.10}$$

$$\frac{d\hat{A}}{dt} = -\hat{k}_{\hat{R}\hat{A}}\hat{R}\hat{A} + \hat{k}_{\hat{P}}\hat{P} + \hat{V}_{\hat{A}}\frac{\hat{P}}{\hat{K}_{\hat{L}} + \hat{P}} + \hat{A}_0 - \hat{k}_{\hat{A}}\hat{A} \tag{3.11}$$

The production of $\hat{R}$ and $\hat{A}$ involve in the process of mRNA transcription. We notice that this process is slow compared to unbinding and binding of $\hat{A}$ and $\hat{R}$ to create complex $\hat{P}$. This way, we consider $\hat{P}$ belongs in quasi steady state

$$\hat{k}_{\hat{P}}\hat{P} = \hat{k}_{\hat{R}\hat{A}}\hat{R}\hat{A}$$

Then, the governing system of equation is as follows,

$$\frac{d\hat{R}}{dt} = -\hat{k}_{\hat{R}}\hat{R} + \hat{V}_{\hat{R}}\frac{\hat{P}}{\hat{K}_{\hat{R}} + \hat{P}} + \hat{R}_0 \tag{3.12}$$

$$\frac{d\hat{A}}{dt} = \hat{V}_{\hat{A}}\frac{\hat{P}}{\hat{K}_{\hat{L}} + \hat{P}} + \hat{A}_0 - \hat{k}_{\hat{A}}\hat{A} \tag{3.13}$$

$$\hat{P} = \frac{\hat{k}_{\hat{R}\hat{A}}}{\hat{k}_{\hat{P}}}\hat{R}\hat{A} \tag{3.14}$$

The quorum sensing is a density dependent behaviour. Let us consider that the autoinducer diffuses bacterial cell membrane and $\hat{\rho}$ is the local cell density and $(1 - \hat{\rho})$ extracellular space (local volume fraction). We assume that quorum sensing molecules (autoinducers) move through cell membrane with $\hat{\delta}$ conductance and $\hat{k}_{\hat{E}}$ natural degrades rate. Suppose, $\hat{E}$ be the concentration of quorum sensing molecule (autoinducer) with uniform cell density and well mixed extracellular space then,

$$(1-\hat{\rho})\left(\frac{d\hat{E}}{dt}+\hat{k}_{\hat{E}}\hat{E}\right)=\hat{\delta}\left(\hat{A}-\hat{E}\right) \tag{3.15}$$

The governing equation for the intercellular autoinducer is

$$\hat{\rho}\left(\frac{d\hat{A}}{dt}-\hat{V}_{\hat{A}}\frac{\hat{P}}{\hat{K}_{\hat{L}}+\hat{P}}-\hat{A}_0+\hat{k}_{\hat{A}}\hat{A}\right)=-\hat{\delta}\left(\hat{A}-\hat{E}\right) \tag{3.16}$$

Finally, the system become

$$\frac{d\hat{R}}{dt}=\hat{V}_{\hat{R}}\frac{\hat{P}}{\hat{K}_{\hat{R}}+\hat{P}}-\hat{k}_{\hat{R}}\hat{R}+\hat{R}_0 \tag{3.17}$$

$$\frac{d\hat{A}}{dt}=\hat{V}_{\hat{A}}\frac{\hat{P}}{\hat{K}_{\hat{A}}+\hat{P}}+\hat{A}_0-\hat{d}\left(\hat{\rho}\right)\hat{A} \tag{3.18}$$

where $\hat{d}(\hat{\rho})=\hat{k}_{\hat{A}}+\frac{\hat{\delta}}{\hat{\rho}}(\frac{\hat{k}_{\hat{E}}(1-\hat{\rho})}{\hat{\delta}+\hat{k}_{\hat{E}}(1-\hat{\rho})})$ and $\hat{P}=\frac{\hat{k}_{\hat{R}\hat{A}}\hat{R}\hat{A}}{\hat{k}_{\hat{P}}}$

We quantitatively analyse the above system and find out stable steady state, a saddle note bifurcation and switching behaviour in the communication system. The switching behaviour is controlled by $\hat{\rho}$.

### 3.3.2 Mathematical Model (PDE)

Next, we can make a more realistic model (i.e. inhomogeneous autoinducer distribution in extracellular space). Suppose, $\hat{L}$ is the thickness of cells (uniform layer). The governing differential equations are

$$\frac{d\hat{A}}{dt}=\hat{F}(\hat{A})+\frac{\hat{\delta}}{\hat{\rho}}(\hat{E}-\hat{A}) \tag{3.19}$$

$$\frac{\partial\hat{E}}{\partial t}=\frac{\partial^2\hat{E}}{\partial x^2}+\frac{\hat{\delta}}{1-\hat{\rho}}(\hat{A}-\hat{E})-\hat{k}_{\hat{E}}\hat{E} \tag{3.20}$$

with Neumann boundary condition $\hat{E}_x(0,t)=0$ and Robin boundary condition $\hat{E}_x(\hat{L},t)+\hat{\alpha}E(\hat{L},t)=0\ (\hat{\alpha}>0)$.

### 3.3.3 Predictions Based on Mathematical Model

Dockery-Keener model is a simple mathematical model of the quorum sensing mechanism (based on known biochemistry) of *P. aeruginosa*. Cell-to-cell communication that works on the elimination of quorum sensing molecule (autoinducer) depends on the density and the colony size. It has been noticed that signalling molecule production turn on to high state as a consequence of autoinducer elimination process which is decreased from extracellular space [3]. So, one notices a switch on/off behaviour in the quorum sensing molecule (autoinducer) production.

## 3.4 Complex Formation Model

Let us consider that the bacterial cell volume is unaltered over times and the model is framed on *P. aeruginosa* quorum sensing molecular mechanism. We consider a single *P. aeruginosa* bacterium emits autoinducers (OdDHL) and changes the extracellular concentration. The mathematical outline can be formed based on the rate of change in concentration of OdDHL (autoinducer), BHL (autoinducer), RsaL, RhlR, LasR and different complexes (RhlR/OdDHL, RhlR/BHL, LasR/OdDHL. The model is based on the following assumptions [4].

- We assume that the BHL, OdDHL, LasR, RasL and RhlR follow Michaelis–Menten kinetics.
- Autoinducer synthesis undergoes without shortage of substrate.
- Post transcriptional activity is not considered in this modelling approach.

Now, we can formulate the rate of change of complexes ($C_1$ for LasR/OdDHL, $C_2$ for RhlR/BHL) and associate $C_3$ for RhlR/OdDHL as

$$\frac{dC_1}{dt} = R_1 A_1 k_1 - C_1 k_2 \tag{3.21}$$

$$\frac{dC_2}{dt} = R_2 A_2 k_3 - C_2 k_4 \tag{3.22}$$

$$\frac{dC_3}{dt} = R_2 A_1 k_5 - C_3 k_6 \tag{3.23}$$

where $k_1$, $k_3$ and $k_5$ are associates and $k_2$, $k_4$ and $k_5$ dissociates rates. $R_1$, $R_2$,$A_1$ and $A_2$ are cellular concentrations of LasR, RhlR, OdDHL and BHL, respectively [4].

Now, we consider that the LasR production is controlled by $K_{R1}$, $V_{R1}$ and $C_1$. $R_{10}$ is an expressed rate of LasR (absence of $C_1$). LasR is degraded with $b_{R1}$ processed rate and LasR is loss with $k_1$ processed rate (due to OdDHL association).

$$\frac{dR_1}{dt} = -R_1 A_1 k_1 + C_1 k_2 - R_1 b_{R1} + V_{R1} \frac{C_1}{K_{R1} + C_1} + R_{10} \tag{3.24}$$

RhlR production is affected by $C_1$ and $V_{R2}$. $K_{R2}$ and $R_{20}$ represent the affinity between *rhlR* and $C_1$ and basal production rate. The dissociation of $C_3$ and $C_2$ increases the concentration of RhlR with rate $k_6$ (for $C_3$) and $k_4$ (for $C_2$). A complex (RhlR/OdDHL) is formed, which corresponds to

$$\frac{dR_2}{dt} = -R_2A_2k_3 + C_2k_4 - R_2A_1k_5 + C_3k_6 - R_2b_{R2} + V_{R2}\frac{C_1}{K_{R2} + C_1} + R_{20} \tag{3.25}$$

where $k_3$ is the associated rate BHL ($A_2$) and RhlR ($R_2$). $k_5$ denotes the associated rate OdDHL ($A_1$) and RhlR.

We can simplify the system (discard the complex formation) as follow,

$$\frac{dR_2}{dt} = -R_2A_2k_3 + C_2k_4 - R_2b_{R2} + V_{R2}\frac{C_1}{K_{R2} + C_1} + R_{20} \tag{3.26}$$

RsaL rate of change is depended on $K_s$, $V_s$ and $C_1$. $S_0$ is the basal production rate of RsaL. We also consider that RsaL is degraded with rate $b_s$.

$$\frac{dS}{dt} = b_sS + V_s\frac{C_1}{K_s + C_1} + S_0 \tag{3.27}$$

Next, the OdDHL production is negatively affected by RasL and positively influenced by the complex $C_1$ (LasR/OdDHL). We assume that RasL works as competitive inhibitor (follow Michaelis–Menten kinetics). OdDHL is produced with maximum rate $V_{A1}$ and $K_{A1}$ denotes the affinity between *lasI* and $C_1$. OdDHL is expressed at rate $A_{10}$ and the affinity between lasI and RasL is represented by $K_{S1}$. We can formulate the rate of production of OdDHL (when the complex is formed with RhlR) as follow [4],

$$\frac{dA_1}{dt} = -R_1A_1k_1 + C_1k_2 - R_2A_1k_5 + C_3k_6 - A_1b_{A1} + V_{A1}\frac{C_1}{C_1 + K_{A1}\left(1 + \frac{S}{K_{S1}}\right)} + A_{10} + d_1(A_{1ex} - A_1) \tag{3.28}$$

where $A_{1ex}$ and $d_1$ are the extracellular concentration and diffusion constant of OdDHL, respectively. When the complex is not formed with RhlR, we get

$$\frac{dA_1}{dt} = -R_1A_1k_1 + C_1k_2 - A_1b_{A1} + V_{A1}\frac{C_1}{C_1 + K_{A1}\left(1 + \frac{S}{K_{S1}}\right)} + A_{10} + d_1(A_{1ex} - A_1) \tag{3.29}$$

Finally, we can formulate the BHL production, which is similar to OdDHL production. In this case, the maximum production rate is $V_{A2}$ and the affinity between *rhlI* promoter and $C_2$ is $K_{A2}$. $A_{20}$ is the BHL expressed rate and $b_{A2}$ is degraded rate of BHL. $A_{2ex}$ parameter represents extracellular BHL concentration. $d_2$ is the diffusion constant for BHL. So, we get

$$\frac{dA_2}{dt} = -R_2A_2k_3 + C_2k_4 - A_2b_{A2} + V_{A2}\frac{C_2}{K_{A2} + C_2} + A_{20} + d_2(A_{2ex} - A_2) \quad (3.30)$$

### 3.4.1 Key Features of the Model

This mathematical framework integrates both quorum sensing systems in *P. aeruginosa*. This model tries to explore the quorum sensing system, when the extracellular concentration of OdDHL is increased. A switching behaviour is observed (uninduced to induced phenotype) due to small changes in extracellular concentration of OdDHL [4]. RsaL plays an important role to hike the extracellular concentration of OdDHL and induced the communication system. On the other hand, Vfr affects the regulatory system strongly and hikes the affinity between *lasR* promoter and the complex LasR/OdDHL.

## 3.5 Dynamical Model of *Staphylococcus Aureus* Cell Communication

*Staphylococcus aureus* is a gram-positive pathogenous bacterium, which is responsible for several diseases. *S. aureus* uses cell-to-cell communication process, which is known as *agr* system. In the communication system virulence genes are expressed in a manner called growth phase dependent. The quorum sensing process is regulated by the autoinducing peptide. One observes that the secretion and the biosynthesis play an important role in experiment. Now, we develop a mathematical system which expresses how *agr* system is influenced by the extracellular concentration of autoinducing peptide. We write the model assumptions as follows [5],

- Degradation is occurred for all substances.
- Synthesis is not delayed.
- We can neglect the stochastic effect because the basel agr P2 promoter pursuit is very high.
- The translation product (level) is proportional to mRNA (levels).
- Transcription follows saturation kinetics and rest of the reaction obeys mass action principle.

- We assume that there are $n$ number of bacterial cells in the cytoplasmic volume $v$. $P$ is concentration of autoinducing peptide and $X$ is inhibiting concentration of autoinducing peptide.
- Substance concentration is homogenous.

Let us assume, $C_{CP}$ and $C_{CX}$ are net amount of complex between AgrC and activating autoinducing peptide and inhibiting autoinducing peptide (in population). $A$, $A_{Pi}$, $C$, $S$ and $R_{III}$ represent net amount of unphosphorylated AgrA, phosphorylated AgrA, non-complex AgrC, SarA and RNAIII in the entire population [5].

In the mathematical modelling approach, we consider that $C_{CP}$ complex (with rate $k_1$) is formed by the $C$ sensor and $P$ autoinducing peptide with $k_2$dissociation rate and $\delta_{CCP}$ degrade rate.

$$\frac{dC_{CP}}{dt} = k_1 CP - k_2 C_{CP} - \delta_{CCP} C_{CP} \tag{3.31}$$

Let, $k_3$ and $k_4$ be the association and dissociation rate in $X$ inhibitory autoinducing peptide and sensor. The complex ($C_{CX}$) is degrade with rate $\delta_{CCX}$

$$\frac{dC_{CX}}{dt} = k_3 CX - k_4 C_{CX} - \delta_{CCX} C_{CX} \tag{3.32}$$

We can formulate rate of RNAIII synthesis with the formalism, $nk_{APi}(\frac{\frac{A_{Pi}}{nv}}{k_{APi}+\frac{A_{Pi}}{nv}})$ $+ nk_B + nk_S(\frac{\frac{S}{nv}}{k_S+\frac{S}{nv}})$ where $\alpha_A$ and $\alpha_C$ are efficiency for translation of agrA and agrC from RNAIII. $k_B$ denote the AgrA-independent activity of P2 (when SarA (S) is absent) [5].

AgrA phosphorylation occurs at the rate $k_p$. Moreover, $\delta_A$ and $\delta_{APi}$ are the degraded rate of AgrA. Now, we assume that $k_{Dp}$ is the rate at which dephosphorylation of AgrA is occurred [5].

$$\frac{dA}{dt} = n\alpha_A k_{APi}\left(\frac{\frac{A_{Pi}}{nv}}{k_{APi} + \frac{A_{Pi}}{nv}}\right) + n\alpha_A k_B + n\alpha_A k_S\left(\frac{\frac{S}{nv}}{k_S + \frac{S}{nv}}\right) - \frac{k_P}{nv} C_{CP} A + k_{DP} A_{Pi} - \delta_A A \tag{3.33}$$

$$\frac{dA_{Pi}}{dt} = \frac{k_P}{nv} C_{CP} A - k_{DP} A_{Pi} - \delta_{APi} A_{Pi} \tag{3.34}$$

AgrC is degraded with the rate $\delta_C$. AgrC and AgA both are synthesized at an equal rate. We consider that the complex (between autoinducing peptide and AgrC) formation occurs at the rate $k_1$. On the other hand, inhibiting autoinducing peptide and AgrC from a complex with rate $k_3$. $k_2$ and $k_4$ represent the dissociation rates of the complexes.

$$\frac{dC}{dt} = n\alpha_C k_{Api}\left(\frac{\frac{A_{pi}}{nv}}{k_{APi} + \frac{Api}{nv}}\right) + n\alpha_C k_B + n\alpha_C k_S\left(\frac{\frac{S}{nv}}{k_s + \frac{S}{nv}}\right) - k_1 CP + k_2 C_{CP} - k_3 CX + k_4 C_{CX} - \delta_C C \tag{3.35}$$

Assume that $\gamma_S$ and $\delta_S$ are constant production rate and the degrade rate of SarA.

$$\frac{dS}{dt} = n_{\gamma S} - \delta_S S \tag{3.36}$$

Then, we get $S = \frac{n_{\gamma S}}{\delta_S}$ at steady state.

RNAIII ($R_{III}$) is degraded at the rate $\delta_{RIII}$. $A_{Pi}$ is required for the transcription of $R_{III}$ and it obeys saturation kinetics with $v_{APi}$ and $k_{APi}$.

$$\frac{dR_{III}}{dt} = nk_{APi}\left(\frac{\frac{A_{Pi}}{nv}}{k_{Api} + \frac{A_{pi}}{nv}}\right) - \delta_{R_{III}} R_{III} \tag{3.37}$$

### 3.5.1 Significance of the Model

The mathematical model of the quorum sensing mechanism of *S. aureus* predicts that the agr-system is hysteretic. The agr-system is triggered at a certain concentration of autoinducing peptide [5]. The system is inactivated for a specific lower autoinducing peptide concentration. Furthermore, the model shows that there is a marginal effect on the RNAIII steady state level.

## References


1. James S, Nilsson P, James G, Kjelleberg S, Fagerström T (2000) Luminescence control in the marine bacterium Vibrio fischeri: an analysis of the dynamics of lux regulation. J Mol Biol 296(4):1127–1137
2. Ward JP, King JR, Koerber AJ, Williams P, Croft JM, Sockett RE (2001) Mathematical modelling of quorum sensing in bacteria. Math Med Biol 18(3):263–292
3. Dockery JD, Keener JP (2001). A mathematical model for quorum sensing in *Pseudomonas aeruginosa*. Bull Math Biol 63(1):95–116
4. Fagerlind MG, Rice SA, Nilsson P, Harlén M, James S, Charlton T, Kjelleberg S (2003) The role of regulators in the expression of quorum-sensing signals in *Pseudomonas aeruginosa*. J Mol Microbiol Biotechnol 6(2):88–100
5. Gustafsson E, Nilsson P, Karlsson S, Arvidson S (2004) Characterizing the dynamics of the quorum-sensing system in *Staphylococcus aureus*. J Mol Microbiol Biotechnol 8(4):232–242

# Mathematical Models of Quorum Sensing Regulated Biofilms Development

4

**Abstract**

Bacterial biofilm is considered as a complex biological phenomenon. The development of the bacterial biofilms is regulated by quorum sensing mechanism. Biofilms have a significant importance in medical science because it is directly related to life-threatening diseases. This chapter deals with the mathematical models of biofilms formation which includes biomass-nutrients model, quorum sensing regulated biofilms models and early development of biofilms.

## 4.1 Biofilms

Bacterial biofilms are the so-called emergent form of bacterial life style, where bacterial cells are embedded in extracellular polymeric substances (EPS). Bacterial cells are aggregated and adhered to each other (or surface) in this complex form of life [1, 2]. Biofilm is completely distinct from the free living bacterial cells. It has emergent properties such as cooperation, competition, localized gradients, sorption, enzyme retention, tolerance and resistance [2]. Biofilms are widely distributed and significant modes of bacterial living culture on earth. The growth and development of biofilms are multistage process (see Fig. 4.1). The planktonic bacteria swim in aqueous medium by their flagella. This swimming process is continuing until they are being attached to sold surface. After this initial attachment, a series of events is initiated. Bacteria lost their flagella and start expanding bacterial colony along the surface. Thereafter, the colony becomes densely packed with bacterial cells. At this stage, the bacterial quorum sensing process regulates biofilm formation and the EPS production begins. Gradually, the volume of the biofilm is enhanced. Finally, biofilm is mature and bacterial cells are embedded by EPS. The heterogenous structure of the biofilm is appeared [3, 4].

Biofilms are found almost everywhere which includes water pipes, on the river bed, in the gut, on the teeth and many more [5]. Biofilms research has great impact


S. Majumdar, S. Roy, *Microbial Communication*,
https://doi.org/10.1007/978-981-15-7417-7_4

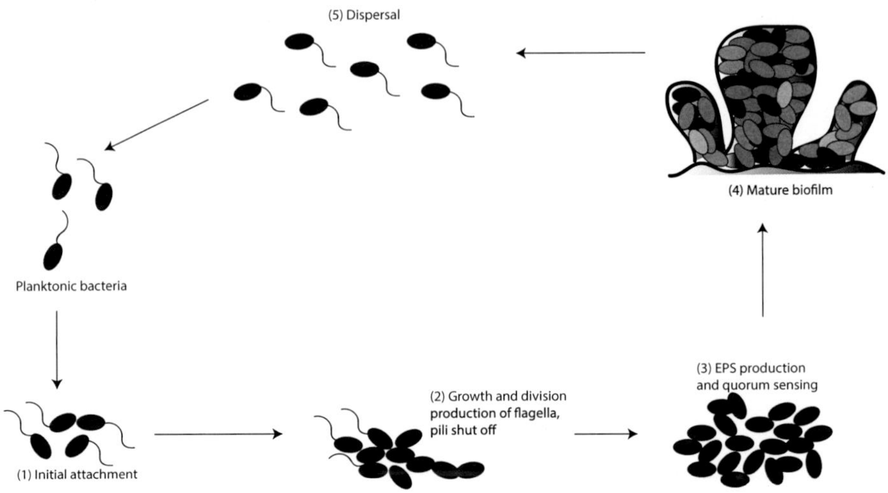


**Fig. 4.1** Schematic diagram of multistage process of bacterial biofilm formation

in medical science, food processing, waste water treatment. Bacterial biofilms are caused by several infection diseases. So, biofilms are clinically significant. Example of biofilms infections in our body include: eye; teeth; urinary tract; lung tissue; gastrointestinal tract and many more [6].

## 4.2 Biomass-Nutrients Model

Bacteria are accumulated inside the biofilm and embedded by the EPS. Bacterial community develops in microcolonies and living in protected growth environment, which involves bacterial infection. The understanding of complex biofilm formation is very important from the point of bacterial infection diseases. Mathematical model of biofilm development has a long history. We begin our journey with a new mathematical framework of spatiotemporal biofilm structure development, with irregular nature [7].

### 4.2.1 Spatiotemporal Model for Biofilm Development

We have assumed several postulates to develop fully continuum model of biofilm growth. All these postulates are nothing but the previous experimental observation and the serious drawback of the previous mathematical approaches. So, we accumulate as follows

- Biomass density cannot cross maximum bound.
- Standard reaction kinetics involve in biomass production.

**Table 4.1** Spatiotemporal model parameters

| Parameter | Description of the parameter |
|---|---|
| $t \geq 0$ | Time (independent variable) |
| $x \in \Omega$ | Space (independent variable) |
| $m(t, x)$ | Biomass density (dependent variable) |
| $c(t, x)$ | Concentration of nutrients |
| $u(t, x)$ | Velocity of flow in liquid region |
| $p(t, x)$ | Pressure of fluid |
| $d_{1,2}(m)$ | Diffusion coefficient for $c$ and $m$ |
| $f(c, m)$ | Nutrient consumption rate |
| $g(c, m)$ | Biomass production rate |

- Compatibility condition: biomass spreading process must be compatible with nutrient transfer and hydrodynamics.
- Spatial heterogeneities in biofilm structure arise due to environmental conditions such as initial and boundary conditions and nutrient availability.
- When a certain maximum density is tensed then biomass spreading has a significant meaning.
- Sharp front of biomass exists at solid/fluid transition.

Biofilm growth model needs accurate description of the environmental condition in liquid region. So, we list related variables in the following Table 4.1 Moreover, $\Omega_1$ (liquid region) and $\Omega_2$ (biofilm structure) are two distinct regions, which are made by biomass density. Now, we get the density dependent biofilm growth model (in general) as,

$$\nabla \cdot u = 0, \ \frac{\partial u}{\partial t} + u \cdot \nabla u = -\frac{1}{\rho}\nabla p + \nabla^2 u \tag{4.1}$$

in $\Omega_1 = \{x \in \Omega \mid m(t, x) = 0\}$, $u \equiv 0$ in $\Omega_2 = \{x \in \Omega \mid m(t, x) > 0\}$

$$\frac{\partial c}{\partial t} + u \cdot \nabla c = \nabla \cdot (d_1(m)\nabla c) - f(c, m) \tag{4.2}$$

$$\frac{\partial m}{\partial t} = \nabla \cdot (d_2(m)\nabla m) + g(c, m) \tag{4.3}$$

$$f(c, m) = \frac{k_1 cm}{k_2 + c}, \ g(c, m) = k_3(f(c, m) - k_4 m) \tag{4.4}$$

with boundary conditions (particular system). Here, fluid flow in the region $\Omega_1$ is described by Eq. 4.1. Nutrient transportation/consumption and biomass density are described by Eqs. 4.2 and 4.3, respectively. Equation 4.4 represents nutrients consumption in region $\Omega_2$ (with $f(c., m)$ reaction rate). It is clear that nutrients are transported in region $\Omega_1$ by convection and diffusion [7]. On the other hand, transportation is diffusive in the region $\Omega_2$. The density dependent diffusion coefficient is described as

$$d_2(m) = \left(\frac{\epsilon}{m_{\max} - m}\right)^a \cdot m^b \tag{4.5}$$

We can physically interpret that $m$ tends to small number means biomass diffusivity is gradually vanished. But, we see this mathematical formalism of biofilm developmental model is little bit complicated to understand. The system is converged to a steady state in two cases, when no nutrients available and no nutrients are added (assume some initial nutrients are available). Now, we can introduce some additional modifications. We assume, $d_1(m) = d_1$ is constant, $C = c/c_0$ and $M = m/m_{\max}$

$$\frac{\partial C}{\partial t} = d_1 \nabla^2 C - F(C, M) \tag{4.6}$$

$$\frac{\partial M}{\partial t} = m_{\max}^{b-a} \nabla \cdot \left[\left(\frac{\epsilon}{1 - M}\right)^a M^b \nabla M\right] + G(C, M) \tag{4.7}$$

with $F(C, M) = K_1 \frac{MC}{K_2 + C}$, $G(C, M) = K_3 \frac{CM}{K_2 + C} - K_4 M$

$$K_1 = m_{\max} \frac{k_1}{c_0},\ K_2 = \frac{k_2}{c_0},\ K_3 = k_3 k_1,\ K_4 = k_3 k_4 \tag{4.8}$$

Finally, we get the above system of reaction–diffusion equation, which can explain the biofilm development. We call it biomass-nutrient model because it shows a spatiotemporal predator–prey model for biomass-nutrients.

### 4.2.2 Spatial Model Predictions

We can simulate the above mathematical model and pop up spatial temporal biofilm structure formation in time. We notice a wavy layer at the beginning, when nutrients are available. The biomass of the system is increasing, when the nutrients concentration is decreasing. Finally, nutrients are limited in the system. Large bacterial colonies start domination over small colonies and mushroom-type structure is visualized (Fig. 4.2).

Biomass-Nutrients model is considered as a biofilm development continuum model. Due to spatial heterogeneities in environmental conditions, the biofilm structure evolves with spatial heterogeneities. This quasilinear reaction–diffusion system of biofilm formation model predicts mushroom-type structure in time, which have a very good agreement with experimental observations [7]. We should think this as basis for biofilm formation model with irregular biofilm structure formation.

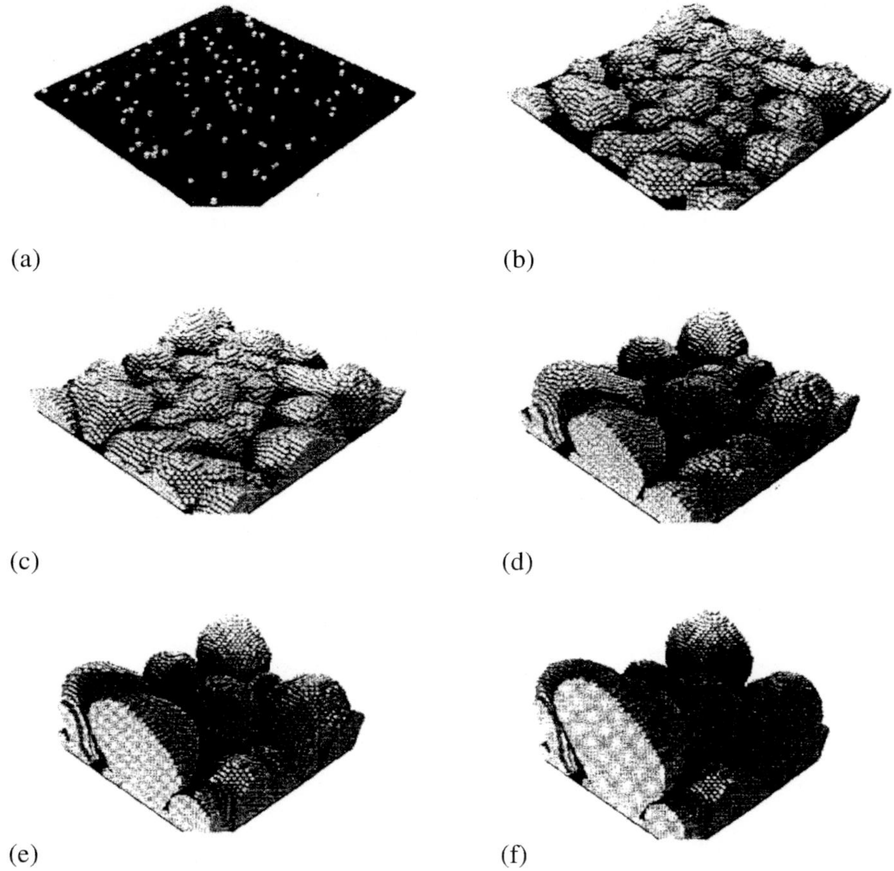


**Fig. 4.2** Biofilm structure development in time. We observe a wavy structure at the beginning. Later bigger colonies dominate the smaller (under nutrient limitation). Smaller colonies grow slower and bigger colonies grow faster (mushroom shapes) (Adapted from [7]). (**a**) $t = 0$.(**b**) $t = 122$ h.(**c**) $t = 235$ h. (**d**) $t = 445$ h. (**e**) $t = 567$ h. (**f**) $t = 643$ h

## 4.3 Quorum Sensing Regulates Biofilm Model

Quorum sensing is a bacterial density dependent behaviour, which controls several phenotypes. Biofilm is one of them, where autoinducers mediated the formation of biofilms with numerous parameters, which include autoinducer diffusion, autoinduction rate, population growth rate and so on. We are now trying to formulate a mathematical approach with monospecies bacterial population inside biofilm [8]. First of all, we list all the parameters for this mathematical model in Table 4.2.

**Table 4.2** Parameters of the model

| Parameter | Description of the parameter |
|---|---|
| $N$ | Bacterial number |
| $N_0$ | Initial bacterial number at time 0 |
| $K$ | Bacterial number in stationary phase |
| $t$ | Time |
| $\rho$ | Growth rate (intrinsic) |
| $\eta_{bc}$ | Autoinducer number inside the cell (calculated in nmol) |
| $\eta_{bf}$ | Autoinducer number inside the biofilm (calculated in nmol) |
| $V_{vc}$ | Total bacterial volume |
| $V_{int}$ | Intercellular space volume of biofilm |
| $V_{bf}$ | Biofilm volume |
| $C_{bc}$ | Autoinducer concentration (inside cell) |
| $C_{bf}$ | Autoinducer concentration (in biofilm) |
| $C_w$ | Autoinducer concentration (in water phase) |
| $h(C_{bc})$ | Autoinducer production rate (net) |
| $B_p$ | Autoinducer production rate (net basal) |
| $\delta$ | Autoinduction rate |
| $A_{bc}$ | Total surface area (for all bacteria) |
| $d$ | Degradation constant |
| $r_{bc}$ | Radius of single bacteria |
| $\gamma$ | This parameter determine how densely the bacteria are packed |
| $\alpha$ | Permeability (cell membrane) |
| $\Delta_{bf}$ | Diffusion rate of autoinducer (out of bacterial cell) |
| $A_{bf}$ | Biofilm surface area |
| $r_{bf}$ | Radius of biofilm |
| $\beta$ | Permeability (surface of biofilm) |
| $\Delta_{fw}$ | Diffusion rate of autoinducer (out of biofilm) |

We consider all bacterial cells are identical and population exhibits logistic growth pattern. So, we model bacterial growth as

$$N(t) = \frac{K}{1 + \left(\frac{K}{N_0} - 1\right) e^{-\rho t}} \tag{4.9}$$

In this approach, autoinducer synthesis (autoinduction) plays a significant role. A positive feedback loop is operated by autoinducer production. Let us consider that autoinducer concentration is proportional to degradation of autoinducers [8]. We have, the net rate of autoinducer production as

$$h(C_{bc}) = B_p - dC_{bc} + g_i(C_{bc}) \tag{4.10}$$

In this scenario, the autoinduction rate is considered to increase monotonically with increasing cellular concentration of autoinducer. It follows,

$$g_{II} = \frac{fC_{bc}}{1 + fhC_{bc}} \tag{4.11}$$

We also assume a negative feedback loop in the system. Thus, the autoinduction production rate is given by

$$g_{log} = \delta C_{bc}\left(1 - \frac{C_{bc}}{C_{bcm}}\right) \tag{4.12}$$

The total surface area for all bacteria is

$$A_{bc} = 4\pi r_b^2 \cdot N(t) \tag{4.13}$$

Accordingly, the concentration of autoinducer inside the bacteria is given by

$$C_{bc} = \frac{\eta_{bc}}{V_{bc}} = \frac{\eta_{bc}}{\frac{4\pi r_{bc}^3}{3} N(t)} \tag{4.14}$$

On the other hand, the autoinducer concentration in the biofilm is

$$C_{bf} = \frac{\eta_{bf}}{V_{int}} \tag{4.15}$$

We have, the intercellular space volume of the biofilm as

$$V_{int} = V_{bf} - V_{bc} \tag{4.16}$$

and the biofilm volume is given by

$$V_{bf} = \gamma V_{bc} \tag{4.17}$$

Now, we calculate the net rate of diffusion of autoinducer out of bacterial cell membranes as follows

$$\Delta_{bc} = \alpha A_{bc}(C_{bf} - C_{bc}) \tag{4.18}$$

We express the rate of change in autoinducer number inside the cell as

$$\frac{d\eta_{bc}}{dt} = h(C_{bc})N(t) - \Delta_{bc} \tag{4.19}$$

Similarly, net rate of diffusion through the biofilm into water is given by

$$\Delta_{bf} = \beta A_{bf}(C_{bf} - C_w) \tag{4.20}$$

Let us consider, the water phase volume is relatively very large (i.e. $C_w \approx 0$). The biofilm surface area is as follows

$$A_{bf} = 4\pi r_{bc}^2 N(t)^{\frac{2}{3}} \gamma^{\frac{2}{3}} \tag{4.21}$$

Moreover, the rate of change in autoinducer number inside the biofilm is expressed as

$$\frac{d\eta_{bf}}{dt} = \Delta_{bc} - \Delta_{bf} \tag{4.22}$$

This model is addressed as quorum sensing regulated phenotypes (i.e. biofilm). Model predicts that a high autoinducer concentration inside cell is reached at low population density. Autoinducer diffusion from biofilm is slower due to extracellular polymeric substance [8]. Moreover, autoinducer regulates biofilms structure formation as well.

## 4.4 Mathematical Model of Bacterial Quorum Sensing and Biofilm

Bacteria are monitoring population density using small molecules (autoinducer) and regulate gene expression accordingly. Now, we present a one-dimensional mathematical model of bacterial communication system inside the growing biofilm. We first address the kinematics of the communication system. Let us consider $f_x$ and $f_w$ be the volume fraction of active biomass and inactive material, respectively [9]. Then $\mu_x f_x \rho_x$ is the net active biomass production.
where

$$\mu_x = Y_{x/o} \frac{\hat{q}_o o}{K_o + o} - b \frac{o}{K_o + o} \tag{4.23}$$

and $o$, $\rho_x$, $Y_{x/o}$, $\hat{q}_o$, $K_o$, $b$ be the concentration of rate limiting substrate (dissolved oxygen), biomass density, yield of active biomass, maximum specific substrate utilization rate, half maximum rate concentration for utilization of substrate, endogenous decay rate coefficient, respectively. $\mu_{wx} f_x \rho_x$ is the production of inactive material.
where

$$\mu_{wx} = (1 - f_D) b \frac{o}{K_o + o} + Y_{w/o} \frac{\hat{q}_o o}{K_o + o} \tag{4.24}$$

Here, $f_D$ and $Y_{w/o}$ are biodegradable fraction of active biomass and yield of EPS, respectively. Oxygen is the limiting substrate, which is consumed with rate $\mu_o f_x \rho_x$. where

$$\mu_0 = -(\hat{q}_o + \gamma f_D b)\frac{o}{K_o + o} \tag{4.25}$$

and $\gamma$ is expressed in mg oxygen/mg VSS. The autoinducer is produced by active biomass with rate $\mu_a f_x \rho_x$.
where

$$\mu_a = \frac{\beta_1 o}{K_0 + o} + \beta_3 + \beta_2 H(a - a_o) \tag{4.26}$$

and $H(x), a, a_o, \beta_1, \beta_2, \beta_3$ be the heaviside function, autoinducer concentration and critical threshold, basel signal production rate (associated with nutrition condition), increased signal production rate in quorum sensing cells, basel signal production rate (not associate with nutrition condition), respectively. The first two terms of the equation express the basal rate of autoinducer production. The quorum sensing is induced when the autoinducer concentration reaches the critical threshold [9]. Finally, the autoinducer (signal) loss coefficient is as follows,

$$\beta_4(pH) = 10^{pH-7} \tag{4.27}$$

where $\beta_4$ be the autoinducer (signal) hydrolysis rate. Thus, the net rate of autoinducer (signal) synthesis is given by $\mu_a f_x \rho x - \beta_4 a$.

Next, we address the biomass equations. We begin with the mass balance equation as follows,

$$\begin{aligned}\frac{\partial}{\partial t}\left[Adz\rho_i f_i(t,z)\right] &= Adz\left(\mu_i(t,z)\rho_i f_i(t,z) + \mu_{ij}\rho_j f_j(t,z)\right) \\ &\quad + Ag_i(t,z) - A\left[g_i(t,z) + \frac{\partial g_i}{\partial z}(t,z)dz\right]\end{aligned} \tag{4.28}$$

Here, $i$,$j$ can be either $x$ or $w$ for active biomass and inactive material, $z$ is the length measured from substratum, $t$ is time, $g_i$ is the mass flux across the surface of small control volume of thickness $dz$, $A$ represents an are orthogonal to the biofilm growth direction. Now, $A$, $\rho_i$ and $dz$ all are taken to be constant. So dividing by them we have

$$\frac{\partial f_i}{\partial t} = \mu_i f_i + \mu_{ij}\frac{\rho_j}{\rho_i} f_j - \frac{1}{\rho_i}\frac{\partial g_i}{\partial z} \tag{4.29}$$

and $u$ is the velocity of the mass at point $z$. So, we get $g_i(t, z) = u(t, z)\rho_i f_i(t, z)$. Substituting the flux into Eq. 4.29 we get

$$\frac{\partial f_i}{\partial t} = \mu_i f_i + \mu_{ij} \frac{\rho_j}{\rho_i} f_j - \frac{\partial}{\partial z}(u f_i) \tag{4.30}$$

So, we can calculate $u$ as follows

$$\frac{\partial u}{\partial z} = \left( \mu_x + \frac{\rho_x}{\rho w} \mu_{wx} \right) f_x \tag{4.31}$$

Please be noted that attachment surface is stationary ($u(t, 0) = 0$). Thus

$$u(t, z) = \int_0^z \left( \mu_x \left(t, z'\right) + \frac{\rho_x}{\rho_w} \mu_{wx} \left(t, z'\right) \right) f_x \left(t, z'\right) dz' \tag{4.32}$$

Here, $\rho_w$ is the inactive material density.

We can determine the change in biofilm thickness as

$$\frac{dL}{dt} = u(t, L) + \sigma(t) \tag{4.33}$$

Finally, we have the evolution equation of the biofilm growth as

$$\frac{\partial f_x}{\partial t} = \mu_x f_x - \frac{\partial}{\partial z}(u f_x) \tag{4.34}$$

Now, we focus on the diffusing quantities. The mass balance for substrate is as follows

$$\frac{\partial o}{\partial t} = \mu_o f_x \rho_x - \frac{\partial J_o}{\partial z} \tag{4.35}$$

where $J_o$ is the flux. This flux can be determined using Fick's Law and biomass velocity:

$$J_o = uo - D_o \frac{\partial o}{\partial z} \tag{4.36}$$

where $D_o$ is the substrate diffusion coefficient in biofilm.

If we substitute this into Eq. 4.35 we have

$$\frac{\partial o}{\partial t} = \mu_o f_x \rho_x - \frac{\partial}{\partial z}(uo) + D_o \frac{\partial^2 o}{\partial z^2} \tag{4.37}$$

The diffusion of signal (autoinducer) is given by

$$\frac{\partial a}{\partial t} = \mu_a f_x \rho_x - \beta_4 a - \frac{\partial}{\partial z}(ua) + D_a \frac{\partial^2 a}{\partial z^2} \tag{4.38}$$

Now, we get a reaction–diffusion system (Eqs. 4.32, 4.33, 4.34, 4.37, 4.38) of biofilm growth coupled with autoinducer (signal) synthesis.

Next, we are rescaling the film coordinates with $\zeta(t) \equiv z/L(t)$, $(0 \le \zeta \le 1)$. We define $\tilde{f}_x(t, z) = f_x(t, \zeta L)$. Hence

$$\frac{\partial}{\partial z} f_x(t, z) = \frac{1}{L} \frac{\partial}{\partial \zeta} \tilde{f}_x(t, \zeta) \tag{4.39}$$

and

$$\frac{\partial}{\partial t} f_x(t, z) = \left( \frac{\partial}{\partial t} - \frac{\zeta u(t, L)}{L} \frac{\partial}{\partial \zeta} \right) \tilde{f}_x(t, \zeta) \tag{4.40}$$

In a similar manner, we can transform for autoinducer (signal) and substrate.

We define $u_L(t) = u(t, L) = \tilde{u}(t, 1)$. After rescaling the coordinates, we obtain final system as

$$\frac{\partial \tilde{f}_x}{\partial t} = \mu_x \tilde{f}_x - \frac{1}{L} \frac{\partial \tilde{u}}{\partial \zeta} \tilde{f}_x + \frac{1}{L}(\zeta u_L - \tilde{u}) \frac{\partial \tilde{f}_x}{\partial \zeta}$$

$$\frac{\partial \tilde{o}}{\partial t} = \mu_o \tilde{f}_x \rho_x - \frac{1}{L} \frac{\partial \tilde{u}}{\partial \zeta} + \frac{1}{L}(\zeta u_L - \tilde{u}) \frac{\partial \tilde{o}}{\partial \zeta} + \frac{D_o}{L^2} \frac{\partial^2 \tilde{o}}{\partial \zeta^2}$$

$$\frac{\partial \tilde{a}}{\partial t} = \mu_a \tilde{f}_x \rho_x - \beta_4 \tilde{a} - \frac{1}{L} \frac{\partial \tilde{u}}{\partial \zeta} \tilde{a} + \frac{1}{L}(\zeta u_L - \tilde{u}) \frac{\partial \tilde{a}}{\partial \zeta} + \frac{D_a}{L^2} \frac{\partial^2 \tilde{a}}{\partial \zeta^2}$$

$$\frac{\partial \tilde{f}_x}{\partial t}(t, 0) = \mu_x \tilde{f}_x - \left( \mu_x + \mu_{wx} \frac{\rho_w}{\rho_x} \right) \tilde{f}_x^2$$

$$\frac{\partial \tilde{f}_x}{\partial t}(t, 1) = \mu_x \tilde{f}_x - \left( \mu_x + \mu_{wx} \frac{\rho_w}{\rho_x} \right) \tilde{f}_x^2$$

$$\frac{\partial \tilde{o}}{\partial \zeta}(t, 0) = 0$$

$$\frac{\partial \tilde{o}}{\partial \zeta}(t, 1) = L J_o \left( o_L - \tilde{o}(t, 1) \right)$$

$$\frac{\partial \tilde{a}}{\partial \zeta}(t, 0) = 0$$

$$\frac{\partial \tilde{a}}{\partial \zeta}(t,1) = LJ_a\left(0 - \tilde{a}(t,1)\right)$$

$$\tilde{u}(t,\zeta) = L\int_0^{\zeta}\left(\mu_x(t,\zeta) + \mu_{wx}(t,\zeta)\frac{\rho_w}{\rho_x}\right)\tilde{f}_x(t,\zeta)d\zeta$$

$$\frac{dL}{dt} = \tilde{u}(t,1) + \sigma(t) \tag{4.41}$$

Here, $\sigma$, $J_a$, $o_L$ are the attachment/detachment rate, signal flux at film surface and substrate concentration in bulk liquid, respectively.

This aforementioned model predicts that there is a correlation between oxygen concentration profile and the biofilm growth. Autoinducer (signal) production is controlled by the bacterial cell of anaerobic region of the biofilm [9]. In the developing biofilm, the quorum sensing is first achieved at the cells near substratum.

## 4.5 Model of Quorum Sensing and Early Development of Biofilms

The mathematical framework of bacterial quorum sensing mechanism is discussed in *lux*- regulation model (see in Chap. 2). Here, we continue the previous lux-regulation model with early development of biofilm. The growth of the biofilm is along with the solid surface. Let $\hat{H}$ be the local hight of the biofilm such that $\hat{H} = \hat{\omega}\hat{N}_T$. $\hat{N}_T$ is the local cross sectional bacterial population density. $\frac{1}{\omega}$ is the bacterial density in the biofilm. Assume $\phi(\hat{N}_T)$ denotes the function of local bacterial cell density which describes the proportion of growth parallel to the surface. A movement is generated by the cell growth in the colony. This movement is described by the velocity $\hat{v}$ tangential to the surface. We assume that the neighbouring bacterial cells occupy the space of death cells. Quorum sensing molecules (autoinducers) are degraded by the biofilm that is mathematically expressed with decay term $\hat{\lambda}$. Suppose, $\hat{A} = \hat{\omega}\hat{N}_T\hat{a}$ or $\hat{A} = \hat{a}\hat{H}$ where $\hat{A}$ is the cross sectional autoinducer concentration. Let us consider $\hat{N}_u$ and $\hat{N}_d$ are up- and down-regulated cross sectional densities, respectively. Furthermore, $\nabla = (\partial/\partial x, \partial/\partial y, 0)$, $x$ and $y$ are the spatial coordinates along biofilm growing surface [10]. Now, the mathematical model of quorum sensing and early development of biofilm is given by

$$\frac{\partial \hat{N}_d}{\partial t} + \nabla\cdot\left(\hat{v}\hat{N}_d\right) = \hat{r}\left(\hat{N}_d + (2-\hat{\gamma})\hat{N}_u\right)\hat{F}\left(\hat{N}_T\right) - \hat{\alpha}\hat{A}\frac{\hat{N}_d}{\hat{\omega}\hat{N}_T} + \hat{\beta}\hat{N}_u \tag{4.42}$$

$$\frac{\partial \hat{N}_u}{\partial t} + \nabla\cdot\left(\hat{v}\hat{N}_u\right) = \hat{r}\left(\hat{\gamma}-1\right)\hat{N}_u\hat{F}\left(\hat{N}_T\right) + \hat{\alpha}\hat{A}\frac{\hat{N}_d}{\hat{\omega}\hat{N}_T} - \hat{\beta}\hat{N}_u \tag{4.43}$$

$$\frac{\partial \hat{A}}{\partial t} + \nabla \cdot \left(\hat{v}\hat{A}\right) = D\nabla^2 \hat{A} + \hat{k}_d \hat{N}_d + \hat{k}_u \hat{N}_u - \hat{A}\left(\hat{\lambda} + \frac{\hat{Q}}{\hat{\omega}\hat{N}_T} + \hat{\alpha}\frac{\hat{N}_d}{\hat{\omega}\hat{N}_T}\right) \tag{4.44}$$

$$\nabla \cdot \hat{v} = \hat{r}\hat{N}_T \phi\left(\hat{N}_T\right)\hat{F}\left(\hat{N}_T\right) \tag{4.45}$$

where $\hat{Q}$ is constant. Adding Eqs. 4.42 and 4.43 we have

$$\frac{\partial \hat{N}_T}{\partial t} + \nabla \cdot \left(\hat{v}.\hat{N}_T\right) = \hat{r}\hat{N}_T \hat{F}\left(\hat{N}_T\right) \tag{4.46}$$

We capture the nutrient limitation effect in the system. $\hat{F}(\hat{N}_T)$ is saturated at $\hat{N}_T = \hat{K}_s$, i.e. $\hat{F}(\hat{K}_s) = 0$. $\hat{K}_s$ is the saturation density per area. Moreover, $\hat{N}_T = \hat{K}_s$ is stable steady state when $\hat{F}'(\hat{K}_s) < 0$. We can express the simplest form (using logistic law) for $\hat{F}(\hat{N}_T)$ as

$$\hat{F}\left(\hat{N}_T\right) = 1 - \frac{\hat{N}_T}{\hat{K}_s} \tag{4.47}$$

Dimensionless model is useful for analysis and future study of the system. We numerically stimulate the model and analyse the long-time behaviour and travelling wave behaviour. A range of travelling wave speed is revealed. This is a basic model of biofilm quorum sensing phenomena where quorum sensing and growth are coupled in the system [10]. The model predictions and experimental observations are quantitatively consistent.

## References

1. Vert M, Doi Y, Hellwich KH, Hess M, Hodge P, Kubisa P, Rinaudo M, Schué F (2012) Terminology for biorelated polymers and applications (IUPAC Recommendations 2012). Pure Appl Chem 84(2):377–410
2. Flemming HC, Wingender J, Szewzyk U, Steinberg P, Rice SA, Kjelleberg S (2016) Biofilms: an emergent form of bacterial life. Nat Rev Microbiol 14(9):563
3. Costerton JW, Cheng KJ, Geesey GG, Ladd TI, Nickel JC, Dasgupta M, Marrie TJ (1987) Bacterial biofilms in nature and disease. Annu Rev Microbiol 41(1):435–464
4. Tolker-Nielsen T, Brinch UC, Ragas PC, Andersen JB, Jacobsen CS, Molin S (2000) Development and dynamics of *Pseudomonas sp.* biofilms. J Bacteriol 182:6482–6489
5. Characklis WG, Marshall KC (1990) Biofilms: a basis for an interdisciplinary approach. In: Characklis WG, Marshall KC (eds) Biofilms. Wiley, New York, pp 3–16
6. Majumdar S, Pal S (2018) Information transmission in microbial and fungal communication: from classical to quantum. J Cell Commun Signal 12(2):491–502
7. Eberl HJ, Parker DF, Van Loosdrecht M (2001) A new deterministic spatio-temporal continuum model for biofilm development. Comput Math Methods Med 3(3):161–175
8. Nilsson P, Olofsson A, Fagerlind M, Fagerström T, Rice S, Kjelleberg S, Steinberg P (2001) Kinetics of the AHL regulatory system in a model biofilm system: how many bacteria constitute a "quorum"?. J Mol Biol 309(3):631–640

9. Chopp DL, Kirisits MJ, Moran B, Parsek MR (2002) A mathematical model of quorum sensing in a growing bacterial biofilm. J Ind Microbiol Biotechnol 29(6):339–346
10. Ward JP, King JR, Koerber AJ, Croft JM, Sockett RE, Williams P (2003) Early development and quorum sensing in bacterial biofilms. J Math Biol 47(1):23–55

# Mathematical Models of Bacterial Swarming Behaviour Regulated by Quorum Sensing

5

**Abstract**

Cell communication process mediates the swarming motility (phenotype) of the bacterial community. Swarming phenomena is triggered by the threshold level of chemical signal and secreted factors. This chapter deals with the mathematical framework of the swarming behaviour, which is regulated by quorum sensing. We emphasize on the agent base mathematical model and find out the active zone, where nutrients, high cell density, signal and secreted factors emerge.

## 5.1 Quorum Sensing and Swarming

Cellular communication process (or quorum sensing) is considered as a gene regulation process, which is controlled by a diffusible chemical signals. Quorum sensing bacterial community adapts several changes in environmental conditions and modifies the defence capabilities. Acylated homoserine lactone is the common chemical signal involved in communication system of gram-negative bacteria. The regulatory network of the communication process contains positive feedback loop (for switch like behaviour). *P. aeruginosa* is a pathogenic gram-negative quorum sensing bacterium, whose swarming behaviour (cells are random motion) is regulated by the quorum sensing molecules (threshold level) [1]. Biosurfactants and enzymes (secreted factors) induce metabolism and the intensity of the movement. Swarming motility is consider as a bacterial community phenomena, where bacterial population move fast on a viscous semisolid surface [1]. Pili and flagella are required for the *P. aeruginosa* swarming behaviour [2]. We characterize the swarming as a connecting movement of bacterial coterie (or group) in a given direction (in the direction of exogenous signals and/or nutrients) [3].

Now, we discuss the computational model.


S. Majumdar, S. Roy, *Microbial Communication*,
https://doi.org/10.1007/978-981-15-7417-7_5

## 5.2 Computational Model of Bacterial Swarming

At first, the agent based model of the complex communication process for bacterial swarming phenomena will be investigated. We consider each bacterial cell is autonomous agent. Bacterial cell regulates their behaviour depending on autoinducers ($S$, $F$) found in environment as well as on the concentration of nutrients. Here, the cells are moving randomly in two-dimensional (2D) plane and interact with each other via autoinducer. Several parameters such as cell movement (swarming states, solitary and activated states) and border movement (constant $k$ and border advancement threshold) play important role. The global parameters in system are medium size $X$ and $Y$, lattice square size, lattice size $X$ and $Y$, initial nutrient per square, initial signal per square and maximum number of bacteria per square [3] (Fig. 5.1).

Let us consider that each autonomous agent follows a simple algorithm (see Fig. 5.2). Cells are regulated functions in a threshold based manner (see Fig. 5.1). Assume, there exist a baseline level of the signal ($S$) production in the planktonic ($P$) or solitary state. The secreted factor ($F$) production begins when the environmental concentration of signal $S$ goes across the threshold level and the production of $S$ grows 5–15-fold. This is called an activated state ($A$). Swarming state ($SW$) has been achieved, when the concentration of $F$ in the environment represses a threshold and as a consequence bacterial cells moving faster and increases the nutrient intake as well. Here, the level of $S$ decreases below threshold, while the level of $F$ remain

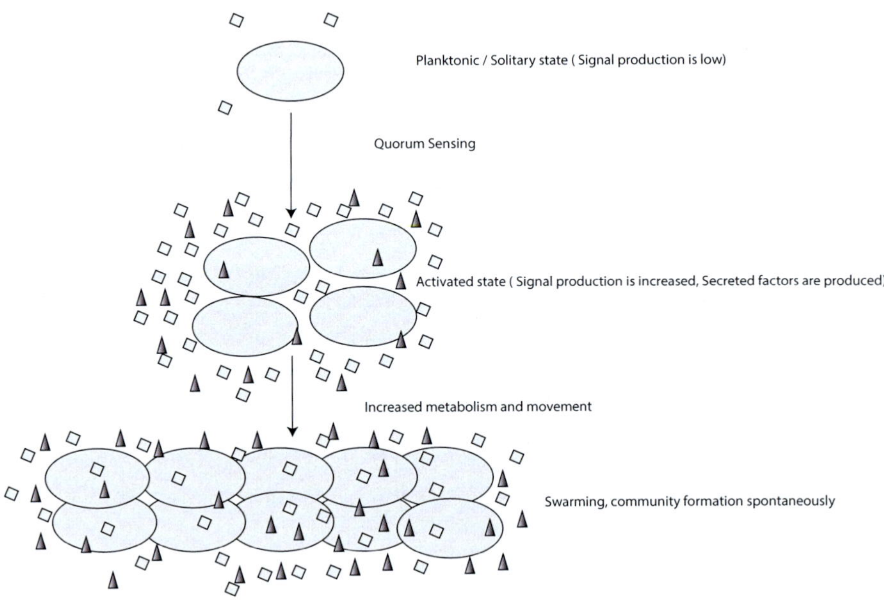


**Fig. 5.1** Schematic diagram of the fundamental principle of quorum sensing regulated swarming in bacteria (*P. aeruginosa*)

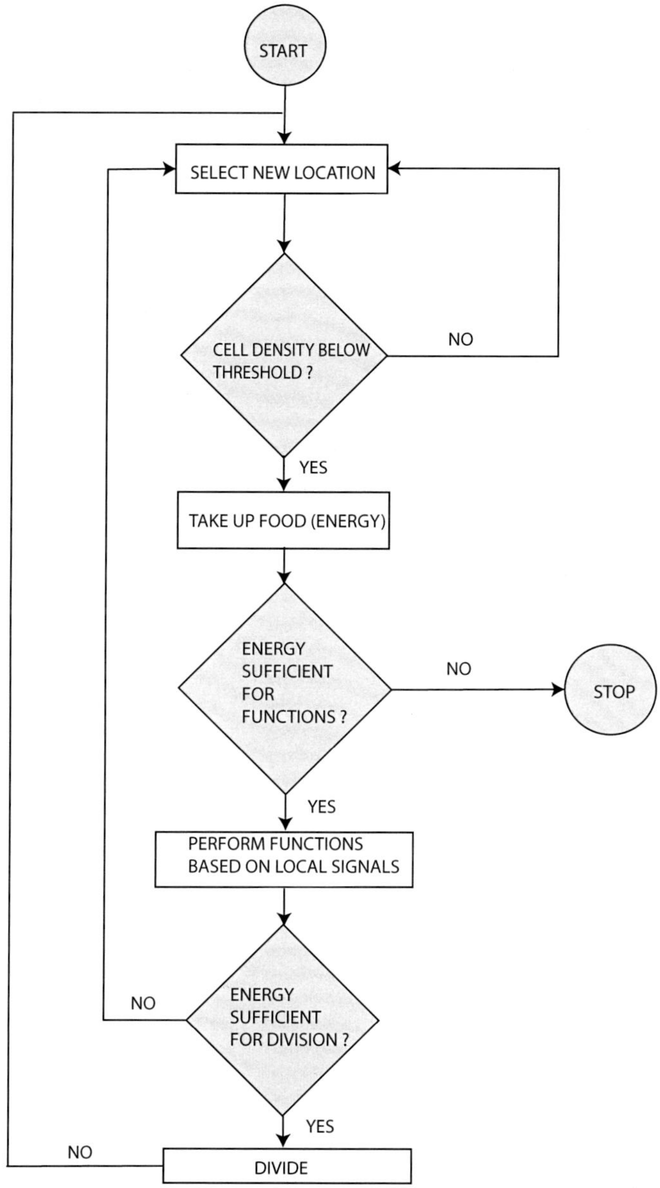


**Fig. 5.2** Agents follow this algorithm at each time point and stop when cells enter into inactive stationary state

the above the threshold. This is the scenario when cells metabolize and move at the rate in swarming state [3].

The energy balance of cells govern the whole process. Cells consume nutrients in a certain amount at each step and spent energy for metabolism, production of $F$ and $S$. Rest of the energy is stored, which is narrated as

$$\tilde{E}(t+1) = \tilde{E}(t) + \tilde{E}(food) - \tilde{E}(S) - \tilde{E}(F) - \tilde{E}(metabolism) \tag{5.1}$$

where $\tilde{E}(t)$ represents a stored energy at time $t$ and the other terms show energy expenditure of nutrient intake $\tilde{E}(food)$, autoinducer signal production ($\tilde{E}(S)$), nutrient production ($\tilde{E}(F)$) and metabolism ($\tilde{E}(metabolism)$). Cell divides when the stored energy increases a threshold, otherwise cell move into a stationary phase. Agents proceed in random steps and in random directions. Cells move three times faster in swarming state than other states [3].

Assume, $N$ denotes all nutrients and the other diffusing materials are $S$ and $F$. The concentration of such component $\tilde{u}$ is expressed by the following equation

$$\frac{d\tilde{u}}{dt} = D\nabla^2\tilde{u} - R\tilde{u} \tag{5.2}$$

where $R$ and $D$ represent uniform decay and diffusion constant, respectively. We further assume that nutrients diffuse only (not decay) and $F$ and $S$ both decay and diffuse.

Moreover, the relative swarming fitness of mutant is calculated as follows

$$SF^{rel} = \frac{p_m \times v_m}{p_{wt} \times v_{wt}} \tag{5.3}$$

where $m$ and $wt$ stand for mutant and wild type, respectively. $p$ and $v$ represent population size and speed of cell in steady state.

Finally, we find out short range relative swarming fitness as follows

$$SF^{rel}_{short\_range} = \frac{p_m}{p_{wt}} \tag{5.4}$$

## 5.3 Significant Model Predictions

The model describes two crucial points. First one is the bacterial colony movement which is density dependent and the second one is displacement towards nutrients (without chemotactic factors). We observe that the bacterial population follow saturation type kinetics and a transition from random population to swarming population (organized). We call this swarming population as an active zone. This active zone is full of nutrients, high bacterial cells number density, secreted factors and signals as well in the concentration, which are required for the swarming behaviour [3]. This zone is asymmetrical from the point of view of distribution of chemical concentration and cell density. Finally, with this mathematical approach it is possible to study quantitatively the early stage of quorum regulated swarming behaviour with the help of speed of advancement and population size.

## References


1. Fraser GM, Hughes C (1999) Swarming motility. Curr Opin Microbiol 2(6):630–635
2. Köhler T, Curty LK, Barja F, Van Delden C, Pechère JC (2000) Swarming of *Pseudomonas aeruginosa* is dependent on cell-to-cell signaling and requires flagella and pili. J Bacteriol 182(21):5990–5996
3. Netotea S, Bertani I, Steindler L, Kerényi Á, Venturi V, Pongor S (2009) A simple model for the early events of quorum sensing in *Pseudomonas aeruginosa*: modeling bacterial swarming as the movement of an "activation zone". Biol Direct 4(1):6

# Mathematical Models of Bacterial Quorum Sensing Regulated Virulence Factors

# 6

**Abstract**

Cell-cell conversation mechanism controls bacterial virulence factors which are considered as sources for different bacterial infection diseases. These pathogenic bacteria are widespread in living systems. Here, we discuss significant mathematical formalism (i.e. deterministic and stochastic model) for the process of endosome escape via virulence factor production in *S. aureus*, competence evoking bacterial communication and quorum sensing inhibitors model.

## 6.1 Deterministic Model

*S. aureus* is a significant human pathogenous Gram-positive bacterium which is associated with several human diseases such as endocarditis, wound infections, osteomyelitis, septic shock, etc. Quorum sensing mechanism controls the virulence factors of *S. aureus* (Fig. 6.1). Here, we deal with the mathematical frameworks for the process of endosome escape via the virulence factor production in *S. aureus* [1]. Our journey begins with deterministic modelling approach. The model is based on some basic assumptions.

- $\tilde{N}_T$ is the total bacterial population and expresses the model for escape from endosome.
- $\tilde{N}_T$ is constant. So, no cell division takes place within endosome.
- Hypothesis: Bacteria are escaped due to accumulation of exo-proteins (i.e. serine proteases, hemolysins, lipase) which are regulated by quorum sensing.
- $\tilde{N}_u$ (up-regulated) and $\tilde{N}_d$ (down-regulated) are two possible bacterial states with $\tilde{N}_T \equiv \tilde{N}_u + \tilde{N}_d$
- $\tilde{a}$ is the concentration of AIP (within endosome).
- $\tilde{c}$ is the concentration of the principal degradative exo-enzyme.
- $\tilde{\rho}$ is the thickness of the endosome membrane.


S. Majumdar, S. Roy, *Microbial Communication*,
https://doi.org/10.1007/978-981-15-7417-7_6

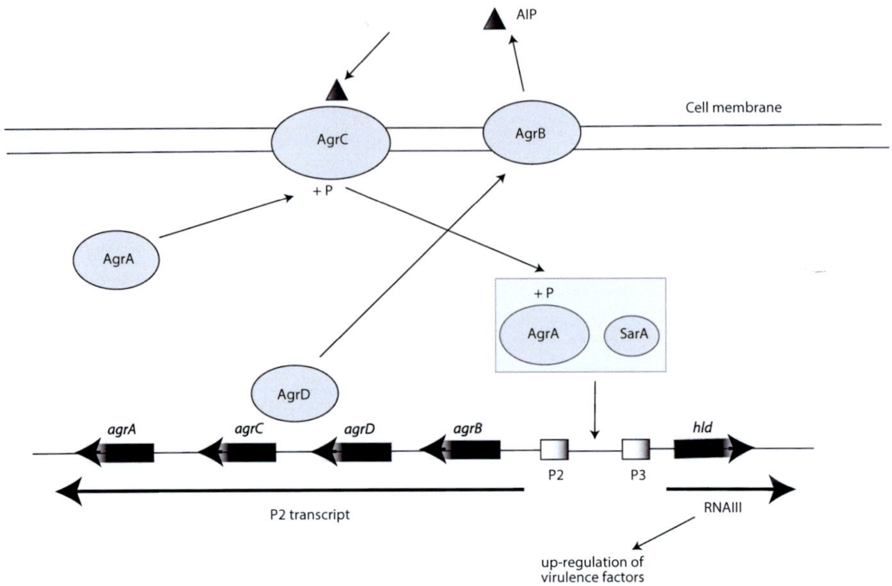


**Fig. 6.1** Schematic diagram of the *agr* system in *S. aureus* (proposed)

- The virulence factors and quorum sensing peptide are produced with low constitutive rate $\tilde{l}_d$ and $\tilde{k}_d$, respectively, at the down-regulated state.
- The virulence factors and autoinducer are generated much faster rate $\tilde{l}_u$ and $\tilde{k}_u$, respectively, at up-regulated state. So we have $\tilde{l}_d \ll \tilde{l}_u$ and $\tilde{k}_d \ll \tilde{k}_u$
- The concentration of autoinducer is proportional to the rate of up-regulation.
- AIP is inactivated at rate $\tilde{\lambda}\tilde{a}$
- $\tilde{\gamma}\tilde{c}$ is the rate of membrane degradation ($\tilde{\gamma}$ is constant).

Thus the deterministic model becomes

$$\frac{d\tilde{N}_d}{dt} = -\tilde{\alpha}\tilde{a}\tilde{N}_d + \tilde{\beta}\tilde{N}_u \tag{6.1}$$

$$\frac{d\tilde{N}_u}{dt} = \tilde{\alpha}\tilde{a}\tilde{N}_d - \tilde{\beta}\tilde{N}_u \tag{6.2}$$

$$\frac{d\tilde{a}}{dt} = \tilde{k}_d\tilde{N}_d + \tilde{k}_u\tilde{N}_u - \tilde{\lambda}\tilde{a} \tag{6.3}$$

$$\frac{d\tilde{c}}{dt} = \tilde{l}_d\tilde{N}_d + \tilde{l}_u\tilde{N}_u \tag{6.4}$$

$$\frac{d\tilde{\rho}}{dt} = -\tilde{\gamma}\tilde{c} \tag{6.5}$$

with initial conditions $\tilde{N}_d(0) = \tilde{N}_T$, $\tilde{N}_u(0) = 0$, $\tilde{a}(0) = 0$, $\tilde{c}(0) = 0$, $\tilde{\rho}(0) = 0$. We can also eliminate $\tilde{N}_u$, using $\tilde{N}_T = \tilde{N}_u + \tilde{N}_d$. The whole system is made non-dimensional using

$$\hat{t} = \frac{\tilde{\alpha}\tilde{k}_u\tilde{N}_T}{\tilde{\beta}}t \tag{6.6}$$

$$\hat{N}_d = \frac{\tilde{N}_d}{\tilde{N}_T} \tag{6.7}$$

$$\hat{N}_u = \frac{\tilde{N}_u}{\tilde{N}_T} \tag{6.8}$$

$$\hat{a} = \frac{\tilde{\alpha}}{\tilde{\beta}}\tilde{a} \tag{6.9}$$

$$\hat{c} = \frac{\tilde{\alpha}\tilde{k}_u}{\tilde{\beta}\tilde{l}_u} \tag{6.10}$$

$$\hat{\rho} = \frac{\tilde{\rho}}{\tilde{\rho}_0} \tag{6.11}$$

We assume $T = \frac{\tilde{\beta}}{\tilde{\alpha}\tilde{k}_u\tilde{N}_T}$ be the characteristic time scale in such a way that the scaling corresponding to equilibrium between both the regulation (up-regulation and down-regulation) [1]. Now, we define constants which are dimensionless.

$$\hat{\lambda} = \frac{\tilde{\lambda}\tilde{\beta}}{\tilde{\alpha}\tilde{k}_u\tilde{N}_T} \tag{6.12}$$

$$k = \frac{\tilde{k}_d}{\tilde{k}_u} \tag{6.13}$$

$$l = \frac{\tilde{l}_d}{\tilde{l}_u} \tag{6.14}$$

$$\epsilon = \frac{\tilde{\alpha}\tilde{k}_u\tilde{N}_T}{\tilde{\beta}^2} \tag{6.15}$$

$$\hat{\gamma} = \frac{\tilde{\gamma}\tilde{l}_u\tilde{\beta}^2}{\tilde{\rho}_0\tilde{\alpha}^2\tilde{k}_u^2\tilde{N}_T} \tag{6.16}$$

Then the mathematical model becomes

$$\epsilon\frac{d\hat{N}_d}{d\hat{t}} = -\hat{a}\hat{N}_d + 1 - \hat{N}_d \tag{6.17}$$

$$\frac{d\hat{a}}{d\hat{t}} = k\hat{N}_d + 1 - \hat{N}_d - \hat{\lambda}\hat{a} \tag{6.18}$$

$$\frac{d\hat{c}}{d\hat{t}} = l\hat{N}_d + 1 - \hat{N}_d \tag{6.19}$$

$$\frac{d\hat{\rho}}{d\hat{t}} = -\hat{\gamma}\hat{c} \tag{6.20}$$

$$\hat{N}_d = \frac{1}{1+\hat{a}} \tag{6.21}$$

The reduced system is as follows,

$$\frac{d\hat{a}}{d\hat{t}} = \frac{k + (1-\hat{\lambda})\hat{a} - \hat{\lambda}\hat{a}^2}{1+\hat{a}} \tag{6.22}$$

$$\frac{d\hat{c}}{d\hat{t}} = \frac{l+\hat{a}}{1+\hat{a}} \tag{6.23}$$

$$\frac{d\hat{\rho}}{d\hat{t}} = -\hat{\gamma}\hat{c} \tag{6.24}$$

subject to $\hat{a} = 0$, $\hat{c} = 0$ and $\hat{\rho} = 0$ at $\hat{t} = 0$. Now, we move to the stochastic modelling approach.

## 6.2 Stochastic Model

We are in a position to formulate the stochastic model of single bacterium from endosome to cytoplasm. Let us assume $\tilde{a}$, $\tilde{c}$, $\tilde{\rho}$ are the concentration of AIP, concentration of degradative exo-enzyme and thickness of the endosome membrane, respectively. Let $P_d$ be the probability density function associated with the down-regulated state. Thus, $P_d(\tilde{a}, \tilde{c}, \tilde{\rho}, t)\Delta\tilde{a}\Delta\tilde{c}\Delta\tilde{\rho}$ is the probability that bacteria are in the down-regulated state (at time $t$). Here, $\tilde{a} \in [\tilde{a}, \tilde{a} + \Delta\tilde{a}]$, $\tilde{c} \in [\tilde{c}, \tilde{c} + \Delta\tilde{c}]$ and $\tilde{\rho} \in [\tilde{\rho}, \tilde{\rho} + \Delta\tilde{\rho}]$. $\Delta$ denotes an infinitesimally small quantity. In a similar fashion $P_u$ is the probability density function related with up-regulation state. Now, $\tilde{\alpha}\tilde{a}\Delta t + o(\Delta t)$ be the probability that down-regulated cell becomes up-regulated cell. Similar way we have, $\tilde{\beta}\Delta t + o(\Delta t)$ as the probability that up-regulated becomes down-regulated cell [1]. Suppose, the local state space volume is changed from time $t$ to $t + \Delta t$. If we assume that cells are down-regulated and $\tilde{a}$ be the amount of autoinducer present at time $t$, then at time $t + \Delta t$ we have

$$\tilde{a} + \tilde{\lambda}\tilde{a}\Delta t - \tilde{k}_d\Delta t + o(\Delta t) \tag{6.25}$$

Here, $\tilde{k}_d \Delta t$ and $\tilde{\lambda}\tilde{a}\Delta t$ are inactivated and produced amount, respectively, by down-regulated cell. If we take a value $\tilde{a} + \Delta\tilde{a}$ at time $t + \Delta t$, then the increased amount can be written as

$$\tilde{a} + \Delta\tilde{a} + \tilde{\lambda}(\tilde{a} + \Delta\tilde{a})\Delta t - \tilde{k}_d \Delta t + o(\Delta t) \tag{6.26}$$

at time t. Subtracting equation 6.25 from 6.26 we get

$$(1 + \tilde{\lambda}\Delta t)\Delta\tilde{a} + o(\Delta t)$$

Next, we calculate the intervals for $\tilde{c}$ and $\tilde{\rho}$. Virulence product concentration at time $t$ corresponding to $\tilde{c}$ at time $t + \Delta t$ is

$$\tilde{c} - \tilde{l}_d \Delta t + o(\Delta t)$$

Here, down-regulated cell produces an amount $\tilde{l}_d \Delta t$ in time $\Delta t$. In the same way we have

$$\tilde{c} + \Delta\tilde{c} - \tilde{l}_d \Delta t + o(\Delta t)$$

is an amount at time $t$ which is corresponding to $\tilde{c} + \Delta\tilde{c}$. Hence this yields a difference $\Delta\tilde{c}$. For $\tilde{\rho}$ we get

$$\tilde{\rho} + \tilde{\gamma}\tilde{c}\Delta t + o(\Delta t)$$

and

$$\tilde{\rho} + \Delta\tilde{\rho} + \tilde{\gamma}\tilde{c}\Delta t + o(\Delta t)$$

So, the interval remains $\Delta\tilde{\rho}$. Similarly, calculation for up-regulated can be done and combining both the cases we get

$$\begin{aligned}
&P_d\,(\tilde{a}, \tilde{c}, \tilde{\rho}, t + \Delta t)\,\Delta\tilde{a}\Delta\tilde{c}\Delta\tilde{\rho} \\
&\quad = P_d\left(\tilde{a} + \tilde{\lambda}\tilde{a}\Delta t - \tilde{k}_d \Delta t, \tilde{c} - \tilde{l}_d \Delta t, \tilde{\rho} + \tilde{\gamma}\tilde{c}\Delta t, t\right)(1 - \tilde{\alpha}\tilde{a}\Delta t) \\
&\qquad \times \left(1 + \tilde{\lambda}\Delta t\right)\Delta\tilde{a}\Delta\tilde{c}\Delta\tilde{\rho} + P_u\left(\tilde{a} + \tilde{\lambda}\tilde{a}\Delta t - \tilde{k}_u \Delta t, \tilde{c} - \tilde{l}_u \Delta t, \tilde{\rho} + \tilde{\gamma}\tilde{c}\Delta t, t\right) \\
&\qquad \times \tilde{\beta}\Delta t\left(1 + \tilde{\lambda}\Delta t\right)\Delta\tilde{a}\Delta\tilde{c}\Delta\tilde{\rho}
\end{aligned} \tag{6.27}$$

Here, down-regulated cell does not become up-regulated with probability $1 - \tilde{\alpha}\tilde{a}\Delta t$ (at time $\Delta t$). Up-regulated cell returns to down-regulated with probability $\tilde{\beta}\Delta t$ (at time $t + \Delta t$). In a same way we have

$$\begin{aligned} & P_u\,(\tilde{a}, \tilde{c}, \tilde{\rho}, t + \Delta t)\,\Delta\tilde{a}\Delta\tilde{c}\Delta\tilde{\rho} \\ &= P_u\left(\tilde{a} + \tilde{\lambda}\tilde{a}\Delta t - \tilde{k}_u\Delta t, \tilde{c} - \tilde{l}_u\Delta t, \tilde{\rho} + \tilde{\gamma}\tilde{c}\Delta t, t\right) \\ &\times \left(1 - \tilde{\beta}\Delta t\right)\left(1 + \tilde{\lambda}\Delta t\right)\Delta\tilde{a}\Delta\tilde{c}\Delta\tilde{\rho} \\ &+ P_d\left(\tilde{a} + \tilde{\lambda}\tilde{a}\Delta t - \tilde{k}_d\Delta t, \tilde{c} - \tilde{l}_d\Delta t, \tilde{\rho} + \tilde{\gamma}\tilde{c}\Delta t, t\right) \\ &\times \tilde{\alpha}\tilde{a}\Delta t\left(1 + \tilde{\lambda}\Delta t\right)\Delta\tilde{a}\Delta\tilde{c}\Delta\tilde{\rho} \end{aligned} \tag{6.28}$$

Here, an up-regulated cell does not become down-regulated with the probability $1 - \tilde{\beta}\Delta t$ (at time $t + \Delta t$). Using Taylor expansion in $\Delta t$ of Eq. 6.27 we express

$$\begin{aligned} & P_d + \Delta t\frac{\partial P_d}{\partial t} + O\left(\Delta t^2\right) \\ &= \left(P_d + \left(\tilde{\lambda}\tilde{a} - \tilde{k}_d\right)\Delta t\frac{\partial P_d}{\partial \tilde{a}} - \tilde{l}_d\Delta t\frac{\partial P_d}{\partial \tilde{c}} + \tilde{\gamma}\tilde{c}\Delta t\frac{\partial P_d}{\partial \tilde{\rho}}\right)\left(1 - \tilde{\alpha}\tilde{a}\Delta t - \tilde{\lambda}\Delta t\right) \\ &+ \left(P_u + \left(\tilde{\lambda}\tilde{a} - \tilde{k}_u\right)\Delta t\frac{\partial P_u}{\partial \tilde{a}} - \tilde{l}_u\Delta t\frac{\partial P_u}{\partial \tilde{c}} + \tilde{\gamma}\tilde{c}\Delta t\frac{\partial P_u}{\partial \tilde{\rho}}\right)\left(\tilde{\beta}\Delta t\right) + O\left(\Delta t^2\right) \end{aligned} \tag{6.29}$$

limit $\Delta t \to 0$ gives

$$\frac{\partial P_d}{\partial t} - \tilde{\lambda}\frac{\partial}{\partial \tilde{a}}(\tilde{a}P_d) + \tilde{k}_d\frac{\partial P_d}{\partial \tilde{a}} + \tilde{l}_d\frac{\partial P_d}{\partial \tilde{c}} - \tilde{\gamma}\tilde{c}\frac{\partial P_d}{\partial \tilde{\rho}} = -\tilde{\alpha}\tilde{a}P_d + \tilde{\beta}P_u \tag{6.30}$$

Similarly, from Eq. 6.28 we get

$$\frac{\partial P_u}{\partial t} - \tilde{\lambda}\frac{\partial}{\partial \tilde{a}}(\tilde{a}P_u) + \tilde{k}_u\frac{\partial P_u}{\partial \tilde{a}} + \tilde{l}_u\frac{\partial P_u}{\partial \tilde{c}} - \tilde{\gamma}\tilde{c}\frac{\partial P_u}{\partial \tilde{\rho}} = \tilde{\alpha}\tilde{a}P_d - \tilde{\beta}P_u \tag{6.31}$$

The above Eqs. 6.30 and 6.31 are coupled system. We analyse the stochastic model with initial conditions. Furthermore, we go forward for non-dimensional system.

The model describes the escape of single bacterium from endosome. We get the asymptotic limit from the model which is biologically relevant. The up- and down-regulation rate are rapid. The system is analysed using steady state analysis and Monte Carlo simulation as well. The endosome escape time [1] can be estimated. We emphasize that this mathematical model shed light on bacterial communication regulated virulence factors as well as several treatment protocols.

## 6.3 Competence Evoking Bacterial Communication

*S. pneumoniae* is a cause of several infection diseases which includes bacteremia, meningitis and respiratory infections. Genetic plasticity is one of the causes for *S. pneumoniae* global success. Quorum sensing regulates competence for genetic transformation. Competence triggers evolution of virulence factors as well as emergence of antibiotic resistance. We are interested to build a mathematical framework of competence evoking system [2]. Several assumptions have been made to simply the model which are described as follows (Fig. 6.2):

- Bacterial population is homogenous.
- The level of mRNA approximate to quasi steady state.
- Competence appears naturally in batch culture.
- Model considers core components.

Let us consider, $X_{ComC}$, $X_{ComD}$, $X_{ComDP}$, $X_{ComE}$, $X_{ComE\sim P}$, $X_{ComX}$, $X_R$ are the total amount of precursor CSP, non-complexed histidine kinase, complexed histidine kinase, response regulator, phosphorylated response regulator, alternative sigma factor and putative repressor, respectively. $X_{CSP}$ is assumed here as the CSP concentration. Now, we give some more details about the parameters used in this mathematical formalism which includes basal synthesis rate ($\beta_{ComC}$ (for ComC), $\beta_{ComD}$ (for ComD), $\beta_{ComE}$ (for ComE)), maximal synthesis rate ($v_{maxComC}$,

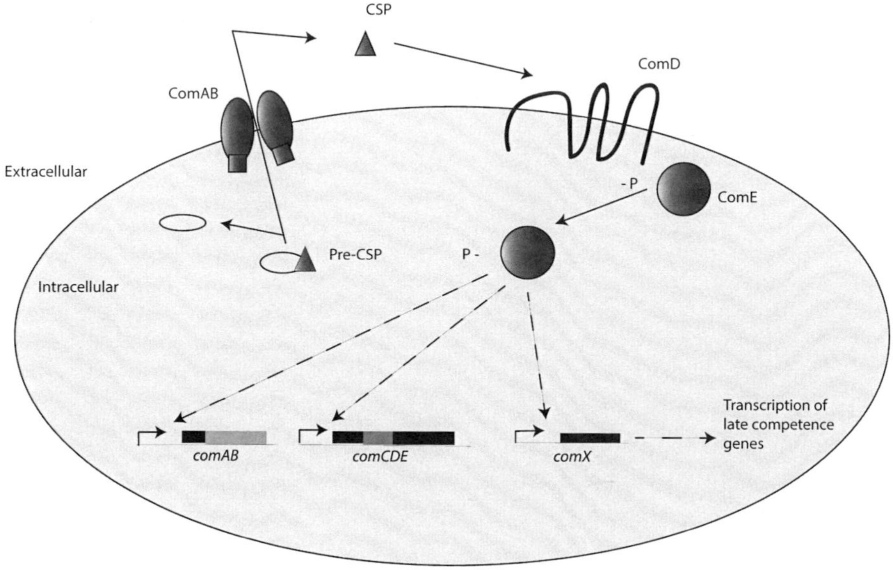


**Fig. 6.2** Schematic diagram of the ComABCDE pathway which is regulated by *S. pneumoniae* quorum sensing system

$v_{maxComD}$, $v_{maxComE}$, $v_{maxComX}$,), required concentration of $ComE \sim P$ for half-maximal synthesis rate of ComC, ComD, ComE, ComX ($k_{ComC}$, $k_{ComD}$, $k_{ComE}$, $k_{ComX}$), extracellular concentration of CSP (CSP), volume of $10^6$ cells ($v_{int}$), decay rate ($\delta_{ComC}$, $\delta_{ComD}$, $\delta_{ComE}$, $\delta_{ComE}$, $\delta_{CSP}$, $\delta_{ComDP}$, $\delta_{ComE\sim P}$, $\delta_{ComX}$), decay rate of putative repressor ($\delta_R$), maximal export rate of ComC ($v_{\max_e xport}$), required concentration of ComC for half maximal export rate ($k_{export}$), complex formation rate of CSP and ComD ($\alpha$), scaling factor ($\psi$), dissociation rate of CSP and ComD ($\gamma$), phosphorylation rate ($\lambda$), dephosphorylation rate ($\rho$), extracellular volume in batch culture ($v_{ext}$), local cell density ($n$), maximal synthesis rate of putative repressor ($v_{maxR}$), required concentration of ComX for half maximal synthesis rate of putative repressor ($k_R$), effect of putative repressor ($\omega$), hill coefficient $\eta$) [2]. We are now in a position to calculate the amount of ComD ($X_{ComD}$) as

$$\frac{dX_{ComD}}{dt} = \beta_{ComD}n + v_{maxComD}n\left(\frac{X_{ComE\sim P}}{X_{ComE\sim P} + nv_{int}k_{ComD}}\right) - \delta_{ComD}X_{ComD} - \alpha X_{ComD}CSP + \gamma X_{ComDP} \tag{6.32}$$

The amount of complex ComD ($X_{ComDP}$) is expressed as

$$\frac{dX_{ComDP}}{dt} = \alpha X_{ComD}CSP - \gamma X_{ComDP} - \delta_{ComDP}X_{ComDP} \tag{6.33}$$

The right-hand side of Eq. 6.33 consists of three term which includes complex formation (first term), complex dissociation (second term) and decay (last term). The changes in total amount of ComE ($X_{ComE}$) and $ComE \sim P$ ($X_{ComE\sim P}$) are expressed as

$$\frac{dX_{ComE}}{dt} = \beta_{ComE}n + v_{maxComE}n\left(\frac{X_{ComE\sim P}}{X_{ComE\sim P} + nv_{int}k_{ComE}}\right) - \delta_{ComE}X_{ComE} - \frac{\lambda}{nv_{int}}X_{ComE}X_{ComDP} + \rho X_{ComE\sim P} \tag{6.34}$$

$$\frac{dX_{ComE\sim P}}{dt} = \frac{\lambda}{nv_{int}}X_{ComDP}X_{ComE} - \rho X_{ComE\sim P} - \delta_{ComE\sim P}X_{ComE\sim P} \tag{6.35}$$

Equation 6.34 is formulated by five terms (i.e. basal synthesis, $ComE \sim P$ induced synthesis, decay, phosphorylation of ComE, dephosphorylation of $ComE \sim P$) and left-hand side of Eq. 6.35 consists of phosphorylation of ComE, dephosphorylation of $ComE \sim P$ and decay terms [2].

Next, we are including CSP synthesis to look at the system behaviour in a growing bacterial cell density. The changes in total amount of intracellular ComC ($X_{ComC}$) are given by

$$\frac{dX_{ComC}}{dt} = \beta_{ComC}n + v_{maxComC}n\left(\frac{X_{ComE\sim P}}{X_{ComE\sim P} + nv_{int}k_{ComC}}\right) - \delta_{ComC}X_{ComC} - v_{\max_e xport}n\left(\frac{X_{ComC}}{X_{ComC} + nv_{int}k_{export}}\right) \tag{6.36}$$

The changes in the concentration of extracellular CSP ($X_{CSP}$) are given by

$$\frac{dX_{CSP}}{dt} = v_{\max_e xport}\frac{n}{v_{ext}}\psi\left(\frac{X_{ComC}}{X_{ComC} + nv_{int}k_{export}}\right) - \alpha\psi X_{CSP}\frac{X_{ComD}}{v_{ext}} + \gamma\psi\frac{X_{ComDP}}{v_{ext}} - \delta_{CSP}X_{CSP} \tag{6.37}$$

So far, we study the initiation of competence. Now we move onto negative feedback because competence is transient state. The amount of ComX ($X_{ComX}$) is given by

$$\frac{dX_{ComX}}{dt} = v_{maxComX}n\left(\frac{X_{ComE\sim P}}{X_{ComE\sim P} + nv_{int}k_{ComX}}\right) - \delta_{ComX}X_{ComX} \tag{6.38}$$

Finally, the total amount of repressor ($X_R$) is expressed as

$$\frac{dX_R}{dt} = v_{maxR}n\left(\frac{X^{\eta}_{ComX}}{X^{\eta}_{ComX} + nv_{int}k^{\eta}_{R}}\right) - \delta_R X_R \tag{6.39}$$

The mathematical formalism describes the competence evoking bacterial communication system, which is focused on molecular mechanism. Model explains the shut-off of *com* system which occurs at transcriptional level. Here, it is important to note that competence system is down-regulated. Putative repressor inhibits the expression of *ComX* and *ComCDE* [2]. This competence is presented in waveform which is supported by experimental evidence.

## 6.4 Quorum Sensing Inhibitors Model

It is well known that many quorum sensing bacteria (i.e. *P. aeruginosa*, *S. aureus*, *V. cholerae*) are human pathogen. So the virulence genes are expressed and control by cell-cell communication process. Thus, quorum sensing inhibitors are possible candidates for effective anti-microbial therapy. Many quorum sensing inhibitors are already been identified. We address combination of inhibitors at molecular level in suppressing virulence. Quorum sensing inhibitors usually apply their effect in

**Table 6.1** List of variables use in the model

| Variables | Descriptions |
|---|---|
| $\tilde{Y}_I$ | Total intracellular concentration of LuxI (enzyme) |
| $\tilde{Y}_R$ | Total intracellular concentration of LuxR (transcription factor) |
| $\tilde{\gamma}_I, \tilde{\gamma}_R$ | Protein degradation rates |
| $\tilde{Q}_I, \tilde{Q}_R$ | Translation rates scaled by respective decay rates |
| $\tilde{\phi}$ | Concentration of AHL |
| $\tilde{k}_{\tilde{\phi}}$ | Specific rate of AHL synthesis by LuxI |
| $\tilde{\gamma}_{\tilde{\phi}}$ | Constant degradation rate |
| $\tilde{Y}_R^*$ | Concentration of LuxR in active form |
| $\tilde{\delta}_0$ | The basal level accounts for week transcriptional activation by LuxR |
| $\tilde{\alpha}$ | Transcription rate |
| $\tilde{R}_c$ | Concentration of AHL analogue (competitive inhibitor) |
| $\tilde{\phi}_0$ | Concentration constant |

different levels such as degradation of AHL, inhibition of AHL synthesis, inhibition of complex formation (AHL- LuxR) and degradation of LuxR. The different inhibition strategies in combination as well as individually will be examined. Now, we move onto derivation of the mathematical formalism of bacterial communication process and keep an eye on the three major dynamics such as cell population, concentrations of intercellular protein and diffusible signal. Let us consider each bacterial cell volume is $\tilde{V}_C$ and bacteria are growing in total volume $\tilde{V}$. $\tilde{\rho}$ is the cell number density and $\tilde{\gamma}_C$ is the constant growth rate. Assume $\tilde{V}_{C\tilde{\rho}}/\tilde{V} \ll 1$ (cell volume fraction) [3]. Hence, we describe the whole system as (Table 6.1)

$$\frac{1}{\tilde{\gamma}_C}\frac{d\tilde{\rho}}{dt} = \tilde{\rho} \tag{6.40}$$

$$\frac{1}{\tilde{\gamma}_I}\frac{d\tilde{Y}_I}{dt} = \tilde{Q}_I \tilde{f}_I(\cdots) - \tilde{Y}_I \tag{6.41}$$

$$\frac{1}{\tilde{\gamma}_R}\frac{d\tilde{Y}_R}{dt} = \tilde{Q}_R \tilde{f}_R(\cdots) - \tilde{Y}_R \tag{6.42}$$

$$\frac{1}{\tilde{\gamma}_{\tilde{\phi}}}\frac{d\tilde{\phi}}{dt} = \frac{\tilde{k}_{\tilde{\phi}}\tilde{V}_C}{\tilde{\gamma}_{\tilde{\phi}}\tilde{V}}\tilde{\rho}\tilde{Y}_I - \tilde{\phi} \tag{6.43}$$

We assume $\tilde{\gamma}_{\tilde{\phi}} \gg \{\tilde{\gamma}_I, \tilde{\gamma}_R\} \gg \tilde{\gamma}_C$ separation of timescale (i.e. cell density level variables are slower than AHL level variables). Considering $\tilde{\gamma}_{\tilde{\phi}} \to \infty$, we have

$$\tilde{\phi}(t) \equiv \tilde{\mu}\tilde{\rho}(t)\tilde{Y}_I(t), \ with \ \tilde{\mu} = \frac{\tilde{k}_{\tilde{\phi}}\tilde{V}_C}{\tilde{\gamma}_{\tilde{\phi}}\tilde{V}} \tag{6.44}$$

The binding of AHL to LuxR is given by

$$\tilde{Y}_R^* = \tilde{Y}_R \left( \tilde{\delta}_0 + \left(1 - \tilde{\delta}_0\right) \frac{\tilde{\phi}^2}{\tilde{K}_{\tilde{\phi}}^2 + \tilde{\phi}^2} \right) \approx \tilde{Y}_R \left( \tilde{\delta}_0 + \left(\tilde{\mu}/\tilde{K}_{\tilde{\phi}}\right)^2 \left(\tilde{\rho}\tilde{Y}_I\right)^2 \right) \tag{6.45}$$

The rate of transcription at virulence promoter is given by

$$\tilde{f}\left(\tilde{\phi}, \tilde{Y}_R\right) = \tilde{\beta} + (1 - \tilde{\beta})(\tilde{Y}_R^*/\tilde{K}_R)^{\tilde{n}}/(1 + (\tilde{Y}_R^*/\tilde{K}_R)^{\tilde{n}}) \tag{6.46}$$

Here, two feedback topologies are considered. We set $\tilde{f}_I = \tilde{f}(\tilde{\phi}, \tilde{Y}_R)$, $\tilde{f}_R = \tilde{\alpha}$ (for LuxI feedback) and $\tilde{f}_R = \tilde{f}(\tilde{\phi}, \tilde{Y}_R)$, $\tilde{f}_I = \tilde{\alpha}$ (for LuxR feedback). Thus we have the dynamics for both feedbacks as

$$\frac{1}{\tilde{\gamma}_R}\frac{d\tilde{Y}_R}{dt} = \tilde{Q}_R\tilde{\alpha} - \tilde{Y}_R \frac{1}{\tilde{\gamma}_I}\frac{d\tilde{Y}_I}{dt} = \tilde{Q}_I\tilde{f}(\tilde{\phi}, \tilde{Y}_R) - \tilde{Y}_I \tag{6.47}$$

$$\frac{1}{\tilde{\gamma}_I}\frac{d\tilde{Y}_I}{dt} = \tilde{Q}_I\tilde{\alpha} - \tilde{Y}_I \frac{1}{\tilde{\gamma}_R}\frac{d\tilde{Y}_R}{dt} = \tilde{Q}_R\tilde{f}(\tilde{\phi}, \tilde{Y}_R) - \tilde{Y}_R \tag{6.48}$$

The expression level of virulence promoter is given by the above equations. We find steady states and quasi steady state solutions of equations 6.47 and 6.48 [3].

The range of inhibitor strategies with respective parameters is stated as

- LuxI competitive inhibitor ($\tilde{\epsilon}_I$)
- LuxI noncompetitive inhibitor ($\tilde{\epsilon}_I$)
- LuxI turnover ($\tilde{\epsilon}_I$)
- LuxR competitive inhibitor (basal induction) ($\tilde{\epsilon}_I = \sqrt{\tilde{\epsilon}_{Rc}}$)
- LuxR competitive inhibitor (moderate induction) ($\tilde{\delta}_1\tilde{\epsilon}_{RC}$)
- LuxR competitive inhibitor (zero induction) ($\tilde{\epsilon}_{Rnc} = \tilde{\epsilon}_{Rc}$)
- LuxR noncompetitive inhibitor ($\tilde{\epsilon}_{Rnc}$)
- LuxR turnover ($\tilde{\epsilon}_{Rnc}$)

Assume $I_{nc}$ be the noncompetitive inhibitor of LuxI. So we have

$$LuxI + I_{nc} \overset{KInc}{\longleftrightarrow} LuxI \cdot I_{nc} \tag{6.49}$$

AHL is assumed to be not synthesized by the bound complex. Then we get the modified version of Eq. 6.44 as

$$\tilde{\phi} = \tilde{\epsilon}_I\tilde{\mu}\tilde{\rho}\tilde{Y}_I \; with \; \tilde{\epsilon}_I \equiv 1/(1 + [I_{nc}]/K_{Inc}) \tag{6.50}$$

Next, $R_{nc}$ be the noncompetitive inhibitor of LuxR. Similarly, we get

$$LuxR + R_{nc} \overset{KRnc}{\longleftrightarrow} LuxR \cdot R_{nc} \tag{6.51}$$

Total LuxR level in Eq. 6.45 is modified by a factor:

$$\tilde{Y}_R \rightarrow \tilde{\epsilon}_{Rnc}\tilde{Y}_R \; with \; \tilde{\epsilon}_{Rnc} \equiv 1/(1 + [R_{nc}]/K_{Rnc}) \tag{6.52}$$

Synthetic AHL analogues are used as competitive inhibitors of LuxR. To model this effect we have the following reactions

$$LuxR + 2AHL \overset{K\tilde{\phi}^2}{\longleftrightarrow} LuxR \cdot AHL_2 \tag{6.53}$$

$$LuxR + 2\tilde{R}_c \overset{K\tilde{R}c^2}{\longleftrightarrow} LuxR \cdot R_{c2} \tag{6.54}$$

The equilibrium concentration of different form of LuxR is expressed as

$$[LuxR]_{free} = \tilde{Y}_R/\left(1 + \tilde{\phi}^2/\tilde{K}^2_{\tilde{\phi}} + [\tilde{R}_c]^2/\tilde{K}^2_{Rc}\right) \tag{6.55}$$

$$[LuxR \cdot R_{c2}] = \tilde{Y}_R\left(\left[\tilde{R}_c\right]^2/\tilde{K}^2_{Rc}\right)/\left(1 + \tilde{\phi}^2/\tilde{K}^2_{\tilde{\phi}} + \left[\tilde{R}_c\right]^2/\tilde{K}^2_{Rc}\right) \tag{6.56}$$

$$[LuxR \cdot AHL_2] = \tilde{Y}_R\left(\tilde{\phi}^2/\tilde{K}^2_{\tilde{\phi}}\right)/\left(1 + \tilde{\phi}^2/\tilde{K}^2_{\tilde{\phi}} + [\tilde{R}_c]^2/\tilde{K}^2_{Rc}\right) \tag{6.57}$$

So, the total active LuxR is given by

$$\tilde{Y}^*_R = \left(\tilde{\delta}_0[LuxR]_{free} + \tilde{\delta}_1[LuxR \cdot R_{c2}] + [LuxR \cdot AHL_2]\right) \tag{6.58}$$

We simplify with the assumption that $\tilde{K}_R$ is the affinity for DNA binding in all forms of active LuxR. Hence

$$\tilde{Y}^*_R \approx \tilde{Y}_R\left(\tilde{\epsilon}_{Rc}\tilde{\delta}_0 + (1 - \tilde{\epsilon}_{Rc})\,\tilde{\delta}_1 + \tilde{\epsilon}_{Rc}\left(\tilde{\phi}/\tilde{K}_{\tilde{\phi}}\right)^2\right) \tag{6.59}$$

where we define

$$\tilde{\epsilon}_{Rc} \equiv 1/\left(1 + [\tilde{R}_c]^2/\tilde{K}^2_{Rc}\right) \tag{6.60}$$

Moreover, we consider that AHL is added externally which is expressed mathematically as

$$\tilde{\phi} = \tilde{\phi}_0 + \tilde{\mu}\tilde{\rho}\tilde{Y}_I \tag{6.61}$$

Finally, using Eqs. 6.50, 6.52, 6.59. 6.61 we get

$$\tilde{Y}_R^* = \tilde{\epsilon}_{Rnc}\tilde{Y}_R\left(\tilde{\epsilon}_{Rc}\tilde{\delta}_0 + (1-\tilde{\epsilon}_{Rc})\,\tilde{\delta}_1 + \tilde{\epsilon}_{Rc}\left(\tilde{\mu}/\tilde{K}_{\tilde{\phi}}\right)^2\left(\tilde{\phi}_0/\tilde{\mu} + \tilde{\epsilon}_I\tilde{\rho}\tilde{Y}_I\right)^2\right) \tag{6.62}$$

The dynamics can be observed by substituting equation 6.62 into Eq. 6.46. The analysis of the mathematical model shows to find that the combination of LuxR noncompetitive inhibitors together with LuxI inhibitors has potential capacity to suppress virulence over a broad range of parameters [3]. Moreover, we have to be very careful for using LuxR competitive inhibitors. This mathematical approach is helpful to design anti-virulence therapies.

## References

1. Koerber AJ, King JR, Williams P (2005) Deterministic and stochastic modelling of endosome escape by *Staphylococcus aureus*: "quorum" sensing by a single bacterium. J Math Biol 50(4):440–488
2. Karlsson D, Karlsson S, Gustafsson E, Normark BH, Nilsson P (2007) Modeling the regulation of the competence-evoking quorum sensing network in *Streptococcus pneumoniae*. Biosystems 90(1):211–223
3. Anand R, Rai N, Thattai M (2013) Interactions among quorum sensing inhibitors. PLoS one 8(4):e62254

# Evolutionary Models of Bacterial Communication Systems

# 7

**Abstract**

Bacteria are communicated using small diffusible chemical signalling molecules. This communication process controls other bacterial social behaviours such as cooperation, conflict and cheating. In this chapter, we model the bacterial diverse and fascinating social lives. The mathematical formalism of evolutionary models deals with bacterial social and cultural behaviours which includes host-manipulation model, two-trait ESS model, quorum sensing evolution in biofilm, cooperation and cheating.

## 7.1 Host-Manipulation Model

Host manipulation is usually used in the context of macroparasites. Here, we use it for all host parasite systems. The fitness of the parasite is related with the neighbours behaviour because of clustered spatial distribution of parasites. Parasite induced damage to the host (virulence) is discussed in the previous chapter. Now we address the fitness function in the case of host-manipulating parasites. The cooperative venture (host manipulation) has a positive fitness impact on the group and the negative fitness impact on individual [1]. We construct a fitness function $W$ (multiplicative form) using group $G$ and individual $I$ component.

$$W(m_1, m_2) = I(m_1)G(n\bar{m}) \tag{7.1}$$

Here, $W(m_1, m_2)$ represents the fitness of an $m_1$ strategist in a group of $m_2$ strategists. This $m$ denotes the individual contribution to the host manipulation, which is scaled between 0 and 1. $I(m_1)$ and $G(n\bar{m})$ are the individual and group


S. Majumdar, S. Roy, *Microbial Communication*,
https://doi.org/10.1007/978-981-15-7417-7_7

fitness function, respectively, ($m_1$ is the declining and $n\bar{m}$ is the rising function). Let us consider, $G$ and $I$ be linear. We get

$$W(m_1, m_2) = (1 - cm_1)(p + n\bar{m}) \tag{7.2}$$

where $p$ and $c$ represent the passive fitness and the cost to manipulation, respectively. We find the evolutionarily stable strategy (ESS) $m(m^*)$ by maximizing $W(m_1, m_2)$ with respect to $m_1$. We set $m_1 = m_2 = m^*$ and solve for $m^*$, we have

$$m^* = \frac{n(d\bar{m}/dm_1) - cp}{cn(1 + (d\bar{m}/dm_1))} \tag{7.3}$$

where the rate of change of mean genotype contribution to the host manipulation within parasite infrapopulation relative to the individual genotype is denoted by $d\bar{m}/dm_1$. The coefficient of relatedness to the group $R$ is expressed as

$$R = \frac{1 + (n-1)r}{n} \tag{7.4}$$

where, r is the pair coefficient of the relatedness and the $n$ is the group size [1].

### 7.1.1 Nonlinear Group Fitness

Now, we move onto the nonlinear group fitness. We assume the group fitness and total group manipulating effort $n\bar{m}$ increases at the same rate. The definite nature of the function $G(n\bar{m})$ depends on the group interaction of investigating biological phenomena. So, realistic group fitness function is expressed as

$$G(n\bar{m}) = 1 - (1 - p)e^{-sn\bar{m}} \tag{7.5}$$

The parameters $s$ and $p$ control the $G(n\bar{m})$ behaviour [1]. Substituting the Eq. 7.5 to 7.1 we get

$$W(m_1, m_2) = (1 - cm_1)\left[1 - (1 - p)e^{-sn\bar{m}}\right] \tag{7.6}$$

Hence, we have

$$m^* = \frac{1}{c} - \frac{1 - (1 - p)e^{-snm^*}}{snR(1 - p)e^{snm^*}} \tag{7.7}$$

### 7.1.2 Sigmoidal Group Fitness

Finally, we can think about sigmoidal group fitness as an alternative. The following function gives us a sigmoidal response, which is scaled between $p$ and 1.

$$G(n\bar{m}) = p + \frac{1-p}{1+100e^{-sn\bar{m}}} \tag{7.8}$$

Substituting Eq. 7.8 into Eq. 7.1, we get

$$W(m_1, m_2) = (1 - cm_1)[p + (1-p)/(1 + 100e^{sn\bar{m}})] \tag{7.9}$$

As a consequence, we have

$$m^* = \frac{1}{c} - \frac{\left(100pce^{-snm^*} + c\right)\left(1 + 100e^{-snm^*}\right)}{(1-p)100Rcsne^{-snm^*}} \tag{7.10}$$

### 7.1.3 Discussion

Host-manipulating model is applicable for the threshold phenomenon such as bacterial collective behaviour (quorum sensing). Cooperative behaviour is very common in bacterial population, where colonies modify their environment in different ways. For example, a pathogenic bacteria *Erwinia carotovora* has bacterial cooperation behaviour. Quorum sensing bacteria use chemical to adjust their levels of cooperation. This cooperative behaviour is switched on when the particular threshold density is achieved. An ESS of manipulating effort for bacterial communication system (quorum sensing) is described by Eqs. 7.7 and 7.10 under certain assumptions such as group size, costless, honest information [1]. This mathematical formalism suggests that extent of host manipulation is outcome of a tension between interhost selection favouring higher and lower manipulating effort.

## 7.2 Two-Trait ESS Model

We are talking about the talking bacteria, with the possibility of reliable quorum-signalling equilibrium. Assume $n$, $s$, $t$ be the colony size, individual's investment in the production of autoinducer (signalling molecule) and total concentration of autoinducer, respectively. $m(t)$ denotes the individual investment in cooperation [2]. In a similar fashion, like the host-manipulating model, we assume

$$w(n, s, m, \bar{m}) = (I(m) \times G(n\bar{m})) - s \tag{7.11}$$

Here, $\bar{m}$ represents the average level of investment in cooperation. $I(m)$ and $G(n\bar{m})$ are individual and group fitness function, respectively.

Let us consider, $m^*(t)$ be the level of investment in cooperation at any given autoinducer concentration and $s^*$ be the level of investment in signalling (at an evolutionary stable quorum-signalling equilibrium). Moreover, $r$ is the coefficient by which group members are related [2]. So, we have

$$\frac{\partial w(n, s, m, \bar{m})}{\partial m} + \left(\frac{1 + r(n-1)}{n}\right) \frac{\partial w(n, s, m, \bar{m})}{\partial \bar{m}} = 0 \; \forall \; n \tag{7.12}$$

and

$$\int f(n) \left[ \frac{\partial w(n, s, m, \bar{m})}{\partial s} + (1 + (n-1)r) m^{*\prime}(ns^*) \times \left( \frac{\partial w(s, m, \bar{m})}{\partial m} + \frac{\partial w(s, m, \bar{m})}{\partial \bar{m}} \right) \right] dn = 0 \tag{7.13}$$

where $f(n)$ is the probability density distribution of the bacterial colony size by individual.

### 7.2.1 Stable Cooperative Effort

Now, we address the stable cooperative effort with two linear functions ($I$ and $G$), for which

$$w(n, s, m, \bar{m}) = (1 - cm)(p + n\bar{m}) - s \tag{7.14}$$

Here, $c$ and $p$ represent the cost of cooperation and passive fitness, respectively. So, we obtain the stable level of cooperative effort as follows

$$m^*(ns^*) = \frac{1 + r(n-1) - cp}{c(1 + r(n-1) + n} \; for \; n \geq \frac{cp - (1-r)}{r}$$

$$m^*(ns^*) = 0 \; for \; n < \frac{cp - (1-r)}{r} \tag{7.15}$$

### 7.2.2 Stable Signalling Effort

Finally, we express the stable level of cooperative effort as follows

$$\int_{n\geq\frac{cp-(1-r)}{r}} f(n)\left[-1+(1+r(n-1))\left(\frac{r(cp+1)+cp-1}{s^{*}c(r(n-1)+n+1)^{2}}\right)\right.$$
$$\left.\times\left(n-cp-2cn\left(\frac{1+r(n-1)-cp}{c(r(n-1)+n+1)}\right)\right)\right]dn$$
$$+\int_{n<\frac{cp-(1-r)}{r}} f(n)[-1]dn=0 \qquad (7.16)$$

$$\Rightarrow s^{*}=\int_{n\geq\frac{cp-(1-r)}{r}} f(n)(1+r(n-1))\left(\frac{r(cp+1)+cp-1}{c(r(n-1)+n+1)^{2}}\right)$$
$$\times\left(n-cp-2cn\left(\frac{1+r(n-1)-cp}{c(r(n-1)+n+1)}\right)\right)dn \qquad (7.17)$$

### 7.2.3 Model Predictions

Two-trait ESS model is a very interesting model of the evolutionary ecology of communication and cooperation in quorum sensing bacteria. Model looks into the individual conflict under this cooperation. When genetic conflict increases, model suggests that group cooperation is depressed and the threshold density for cooperation is increased as well [2]. Conversely, the quorum sensing strength is increased as conflict increases.

## 7.3 Quorum Sensing Evolution in Biofilm

Bacterial communication process and the biofilm formation are interlinked. One of the key features of the biofilm forming bacteria is to secret extracellular polymeric substance (EPS). EPS consists of DNA, smaller amounts of protein and polysaccharide. Finally, it forms a matrix, where bacterial cells are embedded. EPS secretion is controlled by quorum sensing process. We notice this interlink phenomena in several bacteria which include *P. aeruginosa* and *V. cholerae*. The individual based model of biofilm formation can explain the evolutionary nature of various quorum sensing mechanism control EPS production and solve the dual nature of polymer secretion at high cell density [3].

### 7.3.1 Modelling Approach

We focus on pairwise evolutionary competition between three strains such as

- No polymer secretion and no quorum sensing ($EPS^-$)
- Constitutive polymer secretion and no quorum sensing ($EPS^+$)
- Polymer secretion under negative quorum sensing control such that EPS secretion stops at high cell density ($QS^+$)

We simulate the system base of the individual based modelling approach (see in [4]) with associated numerical scheme (see in [5, 6]). The mathematical model express in two-dimensional reaction–diffusion equations with boundary condition. We can simulate the system for the ranges of quorum sensing threshold parameter to examine time timing and the density dependent communication process influence the outcome of the competition ($QS^+$ vs $EPS^+$ and $QS^+$ vs $EPS^-$). We also calculated the evolutionary stable strategy using stability analysis [3].

In a mixed competition between the strain $QS^+$ and $EPS^+$, the both strains behave identically at the initial stage. They all secrete EPS. As the bacterial cells grow and the cell density increases, the autoinducer concentration reaches the threshold level, and $QS^+$ stop polymer secretion. They invest all resources in growth. Substrate availability is high near the upper surface of the biofilm. In this upper surface, $QS^+$ cells achieve burst of cell division (see Fig. 7.1). $EPS^-$ cells are divided rapidly at the beginning of the simulation (without EPS production) and reach higher initial frequency than $QS^+$. $QS^+$ cells face some difficulty at the initial stage because of lower growth rate. Later, $QS^+$ cells take the majority in the biofilm. $QS^+$ strain produces biomass, after suffocating their neighbour ($EPS^-$). This mathematical simulation of quorum sensing phenomena inside biofilm ravel the basic communication, competition interlink among the bacterial groups. We consider this as a basic evolutionary model of bacterial communication system. This model tries to understand, why and how the biofilm formation and quorum sensing process evolve [3]. Moreover, why the slime production of different bacterial species turn on-off in opposite ways.

## 7.4 Cooperation and Cheating

Cooperation is considered as a behaviour, in which other individuals are benefited. But the evolutionary perspective of the cooperation is very difficult to gage because of vulnerability to cheating. Hamilton's inclusive fitness theory gives us an explanation of evolutionary cooperation. Individuals achieve inclusive fitness through direct and indirect fitness effects. We can study the cooperative behaviour in a field of socio-microbiology. Let us consider, bacterial population, in which bacterial cells are generating public good such as exoproducts like virulence, enzymes for food digestion, surfactants. These public good productions are costly

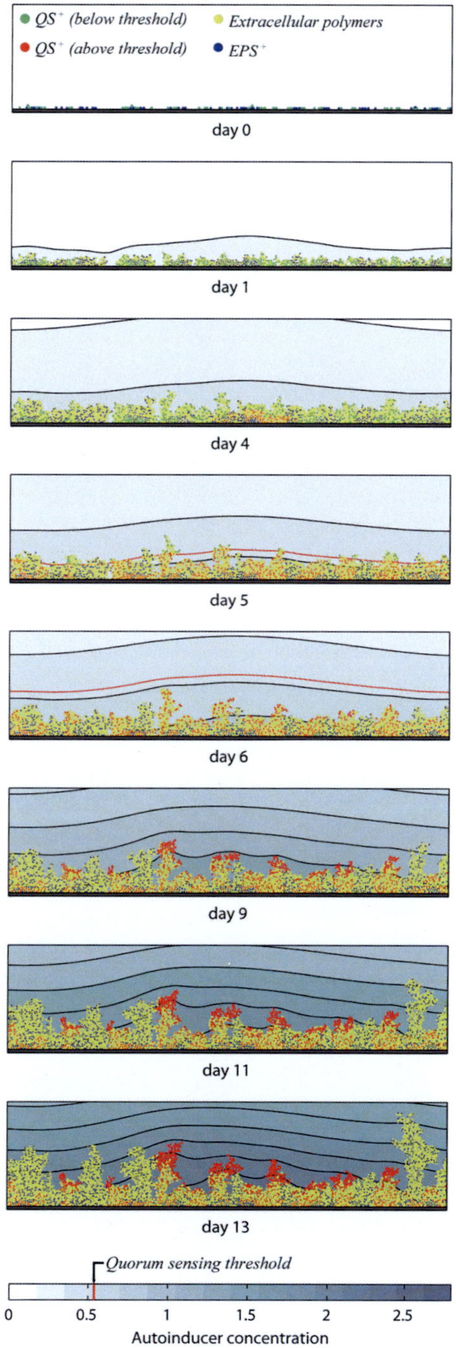


**Fig. 7.1** We can visualize the competition between $QS^+$ and $EPS^+$ with quorum sensing threshold. $QS^+$ and $EPS^+$ behave identically (produce biomass and EPS) until the autoinducer concentration is reached the threshold. After that, $QS^+$ strain stop polymer secretion (Adapted from [3])

as well as beneficial for local group. In the quorum sensing process bacterial collective behaviour is coordinated by small molecules as we already aware of. As a consequence, we observe a coordinated response from the bacterial population and produce some public good. So, one can think, it as a joint activity of collaborators. The evolutionary stability is quite difficult because of cheating activity in the system. As we know, the metabolic cost of signal production is not paid by signal negative strain as well as the cost of responding is not paid by signal-blind strain [7]. By the way, in both cases mutant is benefited by public good production of the surrounding bacteria cells. Now, the question is under what condition the cooperative behaviour breakdown.

### 7.4.1 Framework of Model

Here, we discuss a very general model of quorum sensing evolution using cellular automaton. Quorum sensing control cooperative behaviour of bacteria in the sense of superposition as well as interaction between two cooperative behaviours. The cooperative bacterial communication system, which also coordinates other cooperative behaviour such as public good production. Both types of cooperation are vulnerable to cheating strain. We focus on the parameter combination for which the quorum sensing and cooperation evolve and be maintained [7]. Moreover, how vulnerable the system is for cheating.

Two-dimensional cellular automaton of toroidal lattice topology model is used to simulate the system, where each of $300 \times 300$ grid point represents single bacterium. All the sites are always occupied by bacteria (i.e. no empty sites). We consider three different loci such as **C** locus for cooperation, **S** locus for signal molecule production and **R** locus for signal response. We denote small letter (**c**, **s**, **r**) for inactive allele [7]. Hence, we have eight different genotypes (see Table 7.1) with different metabolic cost and burden.

Now, we focus on the fitness effects of cooperation, where cooperating **C** allele excreted public good. It can increase the fitness of the bacterium (provided at least $n_q$ bacteria expressing **C** within $3 \times 3$ cell neighbourhood). If $n_q$ number of cooperators are present in the neighbourhood, then each individual can get fitness

**Table 7.1** All possible genotypes with phenotypes

| Genotype | Phenotype |
|---|---|
| CSR | Honest |
| CSr | Vain |
| CsR | Shy |
| Csr | Blunt |
| cSR | Lame |
| cSr | Liar |
| csR | Voyeur |
| csr | Ignorant |

benefit from the cooperative behaviour. Cooperator pay a cost (i.e. metabolic burden $M_0$) to enjoy (and/or not enjoy) the benefit of cooperation [7].

Next, we address the fitness effect of quorum sensing. Bacterial cells (genotype **S**) produce quorum sensing molecules (signal molecule) and **R** genotypes is responding in the surrounding environment. So, fitness cost is associated with **S** and **R**. We assume the metabolic fitness cost of quorum sensing and cooperation follows, $m_C >> m_S \geq m_R$. No metabolic cost is associated with inactive alleles. Quorum sensing regulates the cooperation behaviour with $\mathbf{C} \cdot \mathbf{R}$ genome, which express the **C** gene (provided there is $n_q$ individuals in the neighbourhood). On the other hand $\mathbf{C} \cdot \mathbf{r}$ genotypes generate public good.

Now, we move onto selection process, where individuals fight for sites. Competition is carried out between pairs of neighbouring cells, which are selected randomly based on metabolic burdens. The $(M_i)$ net metabolic burden of individual $i$ is depended on basic metabolic load $(M_0)$, $(1 - r)$ unit complement of the cooperation reward parameter and total metabolic cost $(m_e(i))$. Thus,

- $M(i) = [M_0 + m_e(i)]$ if # of cooperators in neighbourhood is below the quorum threshold $n_q$
  $M(i) = [M_0 + m_e(i)](1 - r)$ otherwise $(0 < r < 1)$

So, cooperation reduces the total metabolic burden. The relative fitness of individual is expressed as $M_0/M(i)$. Moreover, diffusion step and mutation is also involve in the simulation process.

### 7.4.2 Results and Conclusions

First, we simulate the system with disabled quorum sensing functions and get two possible genotypes such as Blunt and Ignorant. The left column and the right column of Fig. 7.2 show that the cooperation is costly and cheap, respectively [7]. We summarized the result as follows:

- Cooperation is selected under very low degree of dispersal (in case of $m_c = 30$).
- kin selection plays an important role (in case of $m_c = 30$).
- When the quorum threshold is low then there is a scope of noncooperators to parasitize (In case of $m_c = 30$).
- Cooperation is sustained over a broad range of diffusion rates (in case of $m_c = 10$).
- When the quorum threshold is increasing then the scope for parasitism by noncooperator becomes smaller (In case of $m_c = 10$).

In the end, we simulate the system with quorum sensing and cooperation simultaneously. The model is simulated with high cooperative reward, no diffusion, relatively cheap quorum sensing system, high cost of cooperation and medium quorum threshold. Figure 7.3 gives us the simulation of genotype frequencies,

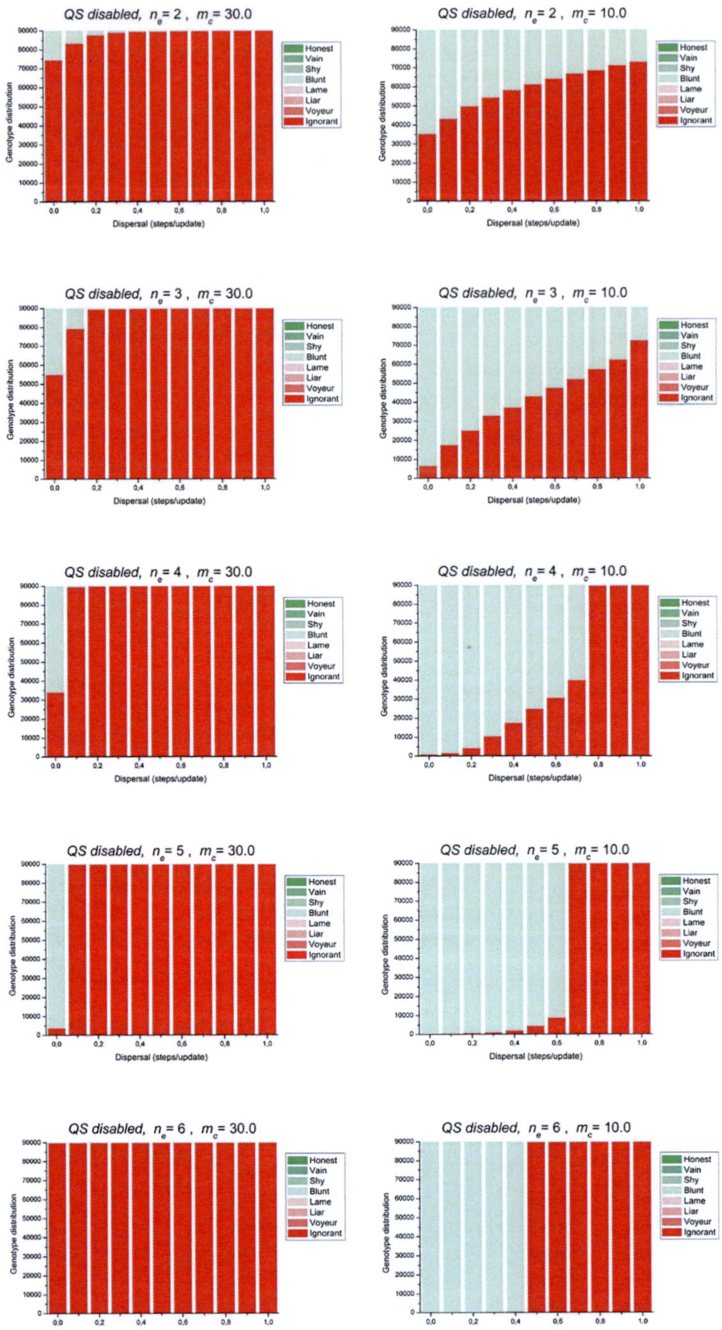


**Fig. 7.2** Stationary genotype distributions of the quorum sensing disabled set of simulations. Here, the fixed parameters are $M_0$ basic metabolic burden, $m_s$ metabolic cost of quorum signal production, $m_r$ metabolic cost of quorum signal response, $r$ fitness reward factor, $\mu_s$, $\mu_r$, $\mu_c$ are mutation rates. Now the screened parameters are $m_c$ metabolic cost of cooperation, $D$ dispersal, $n_e$ quorum threshold (Adapted from [7])

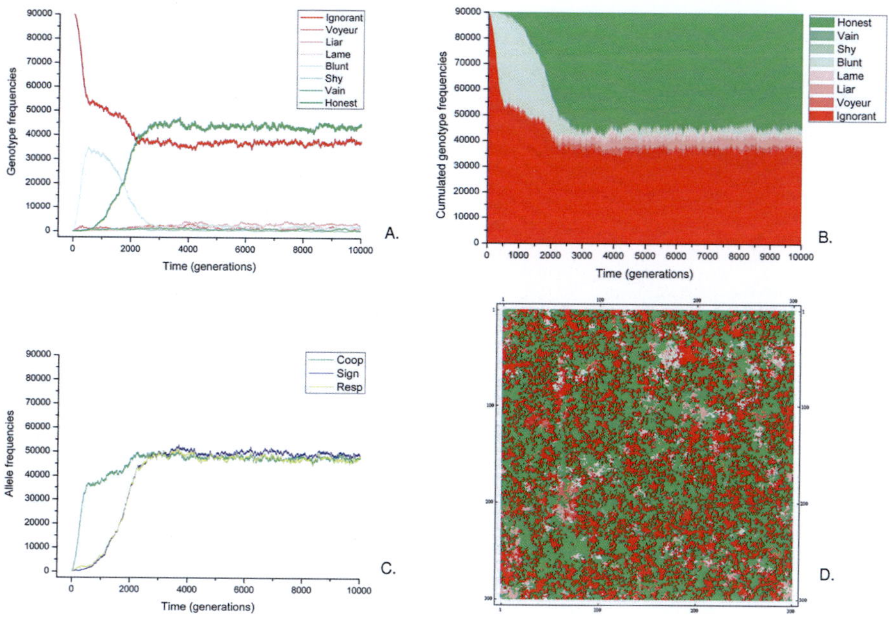


**Fig. 7.3** Details of a single quorum sensing enabled simulation. (**a**) Genotype frequencies, (**b**) Genotype distribution, (**c**) Allele frequencies, (**d**) Spatial pattern of genotypes at $T = 10.0000$ (Adapted from [7])

distribution, allele frequencies and spatial patterns of genotypes. Blunt is the first invading genotype. It lacks quorum sensing and cooperates unconditionally. Honest genotype takes over when Blunt reaches high frequency in population [7]. The majority of the individuals are either Honest or Ignorant. Other genotypes present in low frequencies. We end up with Ignorant and Honest.

Finally, we conclude that quorum sensing regulates the cooperative behaviour and production of public good, with two striking observation such as bacterial communication process enable the cooperative behaviour to attain higher frequency in the population as well as open up rich interaction in which exploitation and cheating share a commonplace.

## References

1. Brown SP (1999) Cooperation and conflict in host–manipulating parasites. Proc R Soc Lond B Biol Sci 266(1431):1899–1904
2. Brown SP, Johnstone RA (2001) Cooperation in the dark: signalling and collective action in quorum-sensing bacteria. Proc R Soc Lond B Biol Sci 268(1470):961–965
3. Nadell CD, Xavier JB, Levin SA, Foster KR (2008) The evolution of quorum sensing in bacterial biofilms. PLoS Biol 6(1):e14

4. Kreft JU, Picioreanu C, Wimpenny JW, van Loosdrecht MC (2001) Individual-based modelling of biofilms. Microbiology 147(11):2897–2912
5. Xavier JB, Picioreanu C, Van Loosdrecht MC (2005) A framework for multidimensional modelling of activity and structure of multispecies biofilms. Environ Microbiol 7(8):1085–1103
6. Picioreanu C, Kreft JU, Van Loosdrecht MC (2004) Particle-based multidimensional multispecies biofilm model. Appl Environ Microbiol 70(5):3024–3040
7. Czárán T, Hoekstra RF (2009) Microbial communication, cooperation and cheating: quorum sensing drives the evolution of cooperation in bacteria. PloS One 4(8):e6655

# Pattern Formation in Bacterial Conversation Mechanisms

# 8

**Abstract**

The bacterial quorum sensing mechanism regulates the gene expression. Autoinducers play a key role to control the system. We emphasise the effective distance range of quorum sensing process, spatiotemporal patterns of gene expression and Turing patterns in gene circuit in this chapter.

## 8.1 Patterns of Quorum Sensing Regulated Gene Expression

Bacterial communication systems (quorum sensing) regulate the gene expression using diffusible signals (autoinducers). This communication process occurs very often in heterogenous environment such as rhizosphere and biofilms. Primarily, autoinducers are not transported via convection and mixing. Thus, autoinducers are transported by diffusion process. The travelling signal can be described in comprehensive manner. We can trigger the fundamental question of effective distance range of the bacterial communication (quorum sensing) by which gene expression is activated. The nonlinearities inherent in gene regulatory mechanism such as promoter activation, signal receptor binding, auto-feedback in signal production have been studied in details [1]. So we expect heterogenous patterns of gene expression, which is developed by autoinducers diffusion across spatially extended bacterial population.

### 8.1.1 Model

Here, we consider LuxI/R system of bioluminescent bacteria called *Vibrio fischeri*. In response to diffusing autoinducer, we formulate a mathematical formalism for the activation of the quorum sensing circuit in the sensor strain (*E. coli* + pJBA132). This mathematical model is based on the multi-well plate data. The formation of


S. Majumdar, S. Roy, *Microbial Communication*,
https://doi.org/10.1007/978-981-15-7417-7_8

green fluorescent protein (GFP) is best for the multistage process. We also consider that three form of concentration which are degraded according to Michaelis–Menten kinetics. The symbols used here as $n$, $\alpha$, $K$, $C$, $D$, $U_1$, $\gamma$, $f$, $a$, $m$, $m_1$, $g(V)$, $k_1$, $k_2$, $U_2$, $m_2$, $G$, $n_0$, $C_\infty$, $L$ and $\nu$ denote cell concentration, intrinsic cell growth rate, cell carrying capacity, autoinducer concentration, autoinducer diffusion constant, concentration of unfolded green fluorescent protein (GFP), proportionality factor, cooperative switch function, half activation constant, Hill coefficient, folding rate of GFP, rate of degradation of GFP in form $V$, maximum degradation rate, Michaelis constant, concentration of folded but non-fluorescent GFP, maturation rate of GFP, concentration of fluorescent GFP, initial cell concentration, fully diffused concentration of autoinducer, length of agar lane and length of the loading region respectively [1].

The above symbols are mathematically expressed as:

$$\frac{dn}{dt} = n\alpha\left(1 - \frac{n}{K}\right) \tag{8.1}$$

$$\frac{\partial C}{\partial t} = D\frac{\partial^2 C}{\partial x^2} \tag{8.2}$$

$$\frac{\partial U_1}{\partial t} = \gamma f(C)\alpha n\left(1 - \frac{n}{K}\right) - m_1 U_1 - g(U_1) \tag{8.3}$$

$$\frac{\partial U_2}{\partial t} = m_1 U_1 - m_2 U_2 - g(U_2) \tag{8.4}$$

$$\frac{\partial G}{\partial t} = m_2 U_2 - g(G) \tag{8.5}$$

$$n(0) = n_0 \tag{8.6}$$

$$C(x,0) = C_\infty L/\nu \quad 0 \le x \le \nu \tag{8.7}$$

$$C(x,0) = 0 \quad \nu < x \le L \tag{8.8}$$

$$U_1(x,0) = U_2(x,0) = G(x,0) = 0 \tag{8.9}$$

$$\frac{\partial C}{\partial x}(0,t) = \frac{\partial C}{\partial x}(L,t) = 0 \quad t > 0 \tag{8.10}$$

where $f(C) = \frac{C^m(x,t)}{a^m + c^m(x,t)}$ and $g(V) = \frac{k_1 V}{k_2 + U_1 + U_2 + G}$.

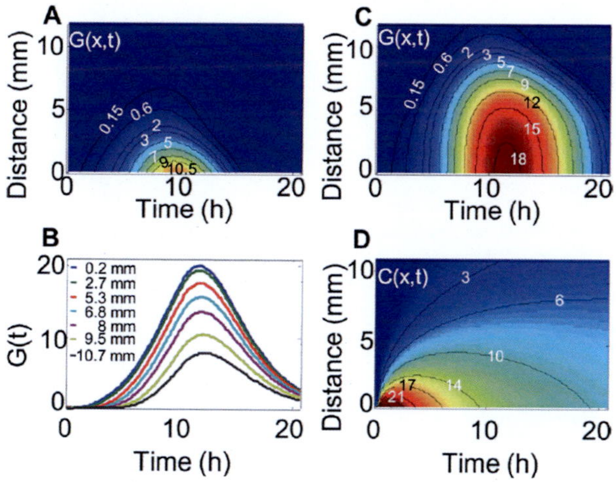


**Fig. 8.1** Patterns are generated by simulating mathematical model. These are patterns of expression predicted for sensor strain in response to diffusing autoinducer. (Reproduce with permission from [1])

### 8.1.2 Spatiotemporal Patterns

The mathematical model is designed on *E.coli* based LuxI/R system which is analog to *V. fischeri*. The sensor strain generates LuxR receptor for autoinducer and responds to exogenous autoinducer by producing GFP. We observe simple pattern using Flick's Law (Eq. 8.2). On the other hand, we simulate the mathematical model and observe spatiotemporal patterns (see Fig. 8.1). $G(x, t)$ evolve for appropriate parameter values for *E. coli* strain. The mathematical simulation has a very good aggregate with the experimental data. The model predicts pattern of activation extending to ~1 cm (for $C_\infty$ = 4 nM autoinducer). Stimulation also matches with experimental data for $C_\infty$ = 0.4 nM autoinducer. The model indicates a simple diffusion of autoinducer (signal) which activates gene expression nonlinearly in bacterial population that flourishes logistically [1]. Further, it is very much sufficient to capture long range synchronized character of patterns.

We conclude that autoinducer (signal) is transported primarily through diffusion, which activate gene expression over macroscopic distance and evolving spatial and temporal patterns (significantly different from simple diffusion). A synchrony is observed on the patterns (length scale at least ~1 cm and time scale ~10 h).

## 8.2 Turing Patterns in Gene Circuit

The ensembles of bacterial cells could self-organize by a well known mechanism called Turing instability. This mechanism leads to spatial periodic patterns. We characterize Turing instability in a gene circuit, which can be implemented in synthetic

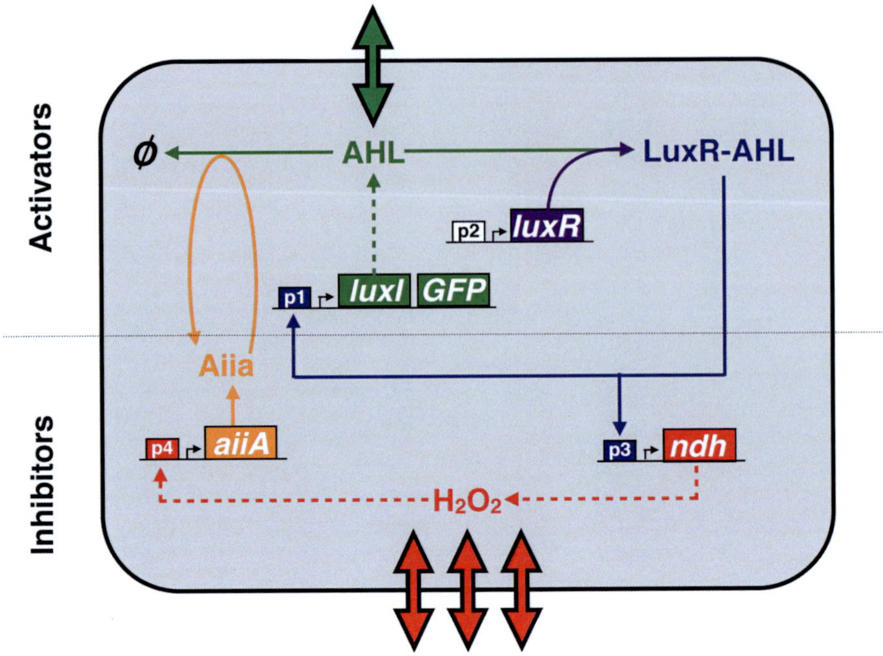


**Fig. 8.2** Schematic diagram of the proposed gene circuit for generating Turing patterns (adapted from [2])

biology as well as population of bacterial cells producing short range activator AHL (N-Acyl homoserine lactone) and long range inhibitor $H_2O_2$ (hydrogen peroxide). The system switches to stable uniform state, limit cycles and Turing patterns. The synthetic gene circuit is designed based on the quorum sensing signalling system. We consider $H_2O_2$ and AHL as long range inhibitor signal and a short range activator signal respectively (see Fig. 8.2). AHL bacterial communication process has native positive feedback, but $H_2O_2$ does not. AHL generates its own production and Aiia degraded it. *ndh* gene produces $H_2O_2$ and $H_2O_2$ generates the transcription of *aiiA* gene. The thick arrow in Fig. 8.2 represents diffusion and intercellular transport of $H_2O_2$ and AHL. The gene circuit is designed by the mathematical model stated below [2]. The compound chemical reaction in the proposed gene circuit is as follows,

$$\emptyset \xrightarrow{\alpha_1 f_1(P)} L \xrightarrow{\gamma_1 g(L)} \emptyset$$

$$\emptyset \xrightarrow{\alpha_2 f_2(P)} H \xrightarrow{\gamma_2} \emptyset$$

$$\emptyset \xrightarrow{\alpha_3 f_3(H)} A \xrightarrow{\gamma_3} \emptyset$$

$$L + L_R \xrightarrow{k_1} P \xrightarrow{k_2} L + L_R$$

where variables $P, L_R, A, H, L$ represent AHL-LUxR complex, LuxR, Aiia, $H_2O_2$ and AHL respectively. We assume that $f_i(Z) = (\delta_i + Z)/(c_i + Z)$ controls the production, and the threshold of production of each species is $c_1 = k_{PL}, c_2 = k_{PH}$, $c_3 = k_{HA}$. Assume $g(L) = A/(k_{LD} + L)$ (enzymatic for AHL). Moreover, we assume that AHL-LuxR complex is relatively stable to degradation and total LuxR is constant. $k_1$ and $k_2$ represent association and dissociation rates respectively [2].

Finally, we have the following set of equations (which include mass-action dynamics of the chemical reaction along with diffusion, $D = \frac{D_H}{D_L}$)

$$\partial_t L = \alpha_1 \frac{\delta_1 + P}{k_{PL} + P} - \gamma_1 A \frac{L}{k_{LD} + L} - k_1(P_m - P)L + k_2 P + \nabla^2 L \tag{8.11}$$

$$\partial_t P = k_1(P_m - P)L - k_2 P \tag{8.12}$$

$$\partial_t H = \alpha_2 \frac{\delta_2 P}{k_{PH} + P} - \gamma_2 H + D\nabla^2 H \tag{8.13}$$

$$\partial_t A = \alpha_3 \frac{\delta_3 + H}{k_{HA} + H} - \gamma_3 A \tag{8.14}$$

where $P_m$ is the total amount of LuxR.

The mathematical framework has been analysed numerically and one gets stable fixed points. We observe that the maximal production rate gives us stable limit cycle oscillation (supercritical Hopf bifurcation), Turing instability and saddle node bifurcation as well (see Fig. 8.3). We find a stable focus ($\alpha = 3.5\,\text{min}^{-1}$), limit cycle ($\alpha_3 = 3.2\,\text{min}^{-1}$), Turing patterns ($\alpha_3 = 2.5\,\text{min}^{-1}$) to numerically simulate the system (in1-D) with periodic boundary conditions. Finally instability leads to form spot patterns (see Fig. 8.4). The model indicates that slowing AiiA feedback on AHL degradation can give us limit cycle oscillation before Turing patterns [2].

Further, it is possible to address the local robustness as well as control the range of unstable wavenumbers (in patterns regime) by tuning signalling parameters. This mathematical approach is considered as a roadmap for optimizing patterns of gene expression. The design is useful for reprogramming spatial dynamics in natural and synthetic gene expression.

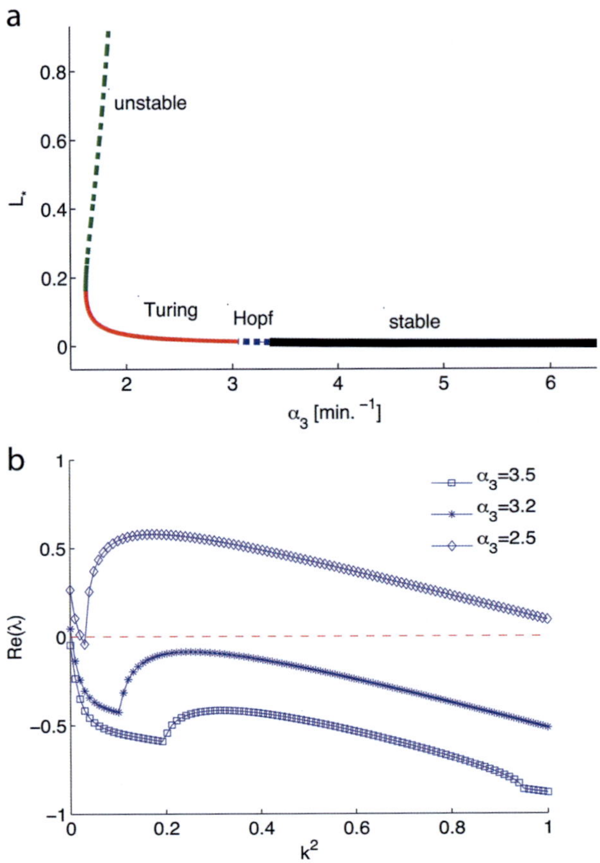


**Fig. 8.3** Slowing Aiia production in mathematical model, which leads to oscillations and turning patterns (adapted from [2])

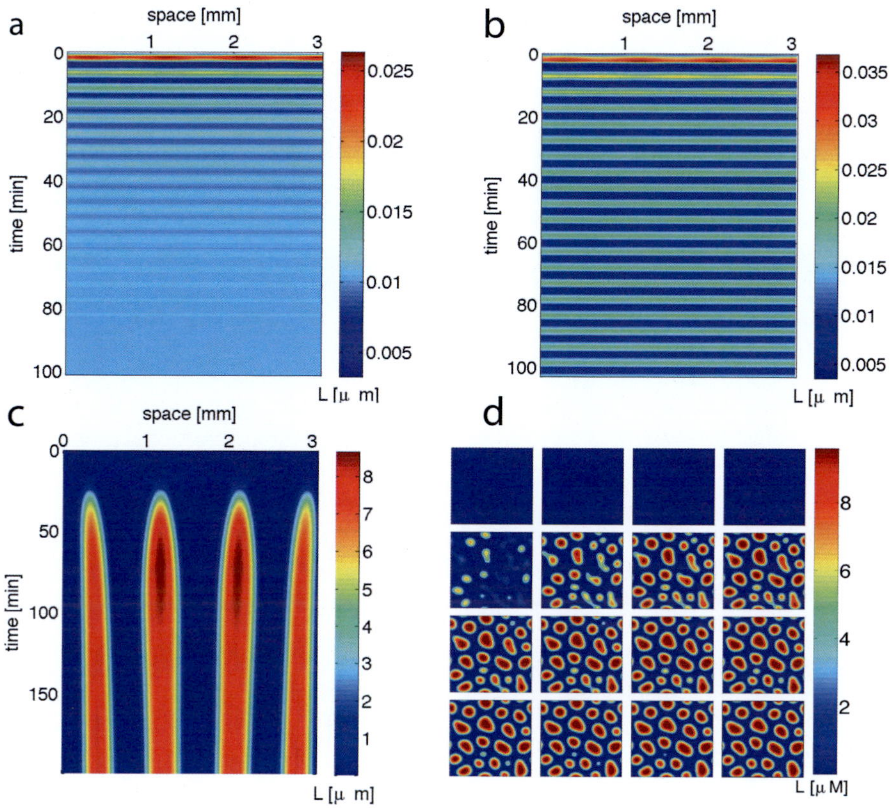


**Fig. 8.4** Numerical simulation of the mathematical model with periodic boundary conditions. Slowing Aiia production confirms (**a**) stable focus ($\alpha = 3.5\,\text{min}^{-1}$), (**b**) limit cycle ($\alpha_3 = 3.2\,\text{min}^{-1}$), (**c**) Turing patterns ($\alpha_3 = 2.5\,\text{min}^{-1}$), (**d**) Spot patterns in two-dimension with periodic boundary conditions and same parameters as (**c**). (Adapted from [2])

## References

1. Dilanji GE, Langebrake JB, De Leenheer P, Hagen SJ (2012) Quorum activation at a distance: spatiotemporal patterns of gene regulation from diffusion of an autoinducer signal. J Am Chem Soc 134(12):5618–5626
2. Borek B, Hasty J, Tsimring L (2016) Turing patterning using gene circuits with gas-induced degradation of quorum sensing molecules. PLoS One 11(5):e0160272

# Summary of Experimental Results

# 9

**Abstract**

In this chapter, we are accumulating significant experimental results of the quorum sensing mechanism over the last few decades. We summarize important experimental observations of several talking bacteria such as *Vibrio fischeri*, *Pseudomonas aeruginosa*, *Agrobacterium tumefaciens*, *Erwinia carotovora*, *Vibrio harveyi*, *Vibrio anguillarum*, *Yersinia enterocolitica*, *Yersinia pseudotuberculosis*, *Aeromonas hydrophila*, *Aeromonas salmonicida*, *Serratia liquefaciens*, *Salmonella typhimurium*, *Ralstonia solanacearum*, *Rhizobium etli*, *Rhodobacter sphaeroides*, *Rhizobium leguminosarum*, *Burkholderia cepacia*, *Chromobacterium violaceum*, *Erwinia chrysanthemi*, *Escherichia coli*, *Vibrio cholerae*, *Streptococcus pneumonic*, *Bacillus subtilis* and *Staphylococcus aureus*.

## 9.1 On *Vibrio fischeri*

- The luminescence genes are triggered by complex (LuxR-autoinducers). *N*-(3-oxohexanoyl) homoserine lactone mediates the *V. fischeri* quorum sensing circuit. Many analogs can be made for this autoinducer. Some analogs can bind with LuxR with same or lower affinity (in case of alteration in acyl side chain). It was noticed that some complex (LuxR-analog autoinducer) had some ability to activate luminescence genes [1].
- LuxI protein is involved in chemical signal synthesis [2].
- We observe two quorum sensing system in *V. fischeri* such as *ain* and *lux*. Luminescence genes expression is triggered by *ain* system at low cell number density. Moreover, *lux* system also regulates luminescence and *ain* system control motility [3].
- S-adenosylmethionine works as amino acid substrate for *V. fischeri* autoinducer synthesis [4].


S. Majumdar, S. Roy, *Microbial Communication*,
https://doi.org/10.1007/978-981-15-7417-7_9

- *V. fischeri* communication system is induced by *N*-phenylacetanoyl-l- homoserine lactones [5].
- *V. fischeri* acetate switch is regulated by AinS quorum sensing. A metabolic relation between cell density and acetate utilization is established [6].
- QsrP, AcfA and RibB production is regulated by LuxR (at transcription level) [7].
- TraR and LuxR bind with autoinducers (3-oxoacyl-HSLs) in a different way, which have a consequence such as a rapid reductions of population density [8].
- Eighteen different genes are controlled by *N*-3-oxohexanoyl-L-homoserine lactone [9].
- *N*-sulfonyl homoserine lactones inhibit the bioluminescence of *V. fischeri* [10].

## 9.2 On *Pseudomonas aeruginosa*

- *Pseudomonas aeruginosa* produces several virulence factors, which are regulated by two quorum sensing circuits (las and rhl). las system controls the rhl system in two levels (transcriptional and posttranslational) [11].
- *P. aeruginosa* is a pathogenic bacteria. The cell-to-cell communication process of *P. aeruginosa* regulates thirty nine genes, which are organized into different classes depending on regulation pattern [12].
- The las and rhl system has transcriptional activator LasI/R and RhlI/R. The las system activates rhlA (mildly) and rhl system activates lasB as well [13].
- 450 and 244 quorum sensing regulated genes are identified at early stationary and mid-logarithmic phase, receptively. Furthermore, quorum sensing control environmental factors such as oxygen availability and medium composition [14].
- The lasI repression by QScR is identified [15].
- Different quorum sensing systems are integrated and make a regulatory networks. Quorum sensing multifunctional single molecules have a significant influence on host–pathogen interaction [16].
- Quorum sensing bacterium *P. aeruginosa* live in biofilms, which is caused for cystic fibrosis lungs [17].
- *P. aeruginosa* quorum sensing circuit is inhibited by azithromycin [18].
- Pyocyanin is a signal, which controls quorum sensing genes [19].
- Bacteria exhibit different social behaviours. Among them, bacterial communication process and biofilm formation are connected. Moreover, researchers introduce a new term called sociomicrobiology [20].
- RpoS controls virulence factors and the communication process of *P. aeruginosa* [21].
- Gene expression is regulated by several families of bacterial communication modulons (in case of *P. aeruginosa*) [22].
- The compound meta-bromo-thiolactone is a quorum sensing inhibitor (QSI), which blocks the *P. aeruginosa* virulence gene expression as well as biofilm formation [23].

- RhlR is the most important target for the therapeutic intervention of infections [24].
- The experiment demonstrated a cooperative activity in bacteria, which is regulated by quorum sensing. This phenomena in *P. aeruginosa* is known as policing process (social cheater) [25].
- We detect communication signalling metabolites in a developing biofilms of *P. aeruginosa* by Raman scattering spectroscopy [26].
- MvfR controls several virulence factors and it is an important anti-virulent target as well [27].
- PqsR, PqsE and 2-alkyl-4-quinolones are the key component of the pqr communication system of *P. aeruginosa*, which activates the virulence genes and biofilm development [28].
- Mycofabricated silver nanoparticles have anti-virulence property, which inhibits production of virulence factors, biofilm formation and cell-to-cell communication process of *P. aeruginosa* [29].
- Bacterial spatial distribution and phenotypic heterogeneity in signals give us new light in the field ecological studies [30].

## 9.3 On *Agrobacterium tumefaciens*

- *Agrobacterium tumefaciens* has multicomponent signal turnover communication system, which is important for genetic regulatory mechanism [31].
- GABA controls the quorum sensing signals of *A. tumefaciens*. GABA is also a signal between plant and bacteria [32].
- Quorum sensing regulated surface motility is an important factor in a microbial competition (in a dual species community). *A. tumefaciens* is emigrated from biofilms (coculture) [33].
- Ti plasmid encoded protein inhibits TraR activity [34].
- Autoinducers of *A. tumefaciens* are detected by the ultrasensitive bioassay system [35].
- Autoinducers are degraded by some proteins (quorum quenching enzymes). These proteins can block several quorum sensing regulated phenomena. AiiB is a quorum quenching lactones form *A. tumefaciens* [36].
- Pyranose-2-phosphate is an identified molecular motif for quorum sensing mechanism [37].
- Go/No go is a system, which controls the quorum sensing and quenching in pathogenic *A. tumefaciens* [38].
- Azamacrolides are a quorum sensing inhibitors for the *A. tumefaciens* [39].
- T4SS and T6SS activity is inactivated by cyclic di-GMP [40].

## 9.4 On *Erwinia carotovora*

- *Erwinia carotovora* is a quorum sensing bacteria. Cell communication process regulates the production of carbapenem antibiotic and virulence factors. We observe a nonenzymatic turnover in quorum sensing molecule [41].
- VirR regulates the virulence in *E. carotovora* [42].
- Quorum sensing mechanism of *E. carotovora* is controlled by the cooperation of two distinct ExpR ( ExpR1 and ExpR2) [43].
- Ten essential oil components are identified as quorum sensing inhibitors for *E. carotovora* [44].
- Hexanal inhibits *E. carotovora* biofilm formations [45].

## 9.5 On *Vibrio harveyi*

- *Vibrio harveyi* is a marine bacteria, which has two quorum sensing process regulated by autoinducer-1 and autoinducer-2 [46].
- Together with two quorum sensing process, there is third way of quorum sensing system. Third quorum sensing process is identified by the genetic analysis of *V. harveti* (*cqsS* and *cqsA*) [47].
- Autoinducer-1 and autoinducer-2 are detected by LuxN and LuxPQ, respectively. Moreover, coincidence detector for both autoinducers regulates gene expression [48].
- Cell-to-cell communication process represses type III secretion in *V. harveyi*, at high cell density [49].
- The luminescence is controlled by LuxO with $\sigma^{54}$ [50].
- The regulatory components are shared by the different quorum sensing process in *V. harveyi* [51].
- Osmotic stress response is regulated by quorum sensing mechanism [52].
- Proteome dynamics is characterized by BONCAT. One hundred and seventy six proteins are identified in different transition stages [53].
- Siderophore production is regulated by quorum sensing mechanism of *V. harveyi* [54].
- *V. harveyi* biofilms are inhibited by undecanoic acid (with auxins) [55].

## 9.6 On *Vibrio anguillarum*

- *N*-(3-oxodecanoyl)-L-homoserine lactone is an autoinducer, which regulates quorum sensing system (VanI/R) of *Vibrio anguillarum* [56].
- T6SSs are very important for virulence activity in gram-negative bacteria *V. anguillarum*. T6SS also regulates PrtV, EmpA and general stress response as well [57].
- vanT is stabilized by sigma factor of RpoS. RpoS is working with quorum sensing circuit for survival and regulates stress responses [58].

- *V. anguillarum* homologues of LuxR, LuxMN, LuxPQ, LuxOU are VanT, VanMN, VanPQ, VanOU, respectively [59].
- Recent result demonstrates antiphage defence strategy link with quorum sensing mechanism in *V. anguillarum* [60].

## 9.7 On *Yersinia enterocolitica*

- Cell-to-cell communication process of *Y. enterocolitica* regulates swarming motility and swimming [61].
- O-glycosylated flavanones (orange extract) is a quorum sensing inhibitor. It inhibits biofilm formation, swimming and swarming motility of *Y. enterocolitica* [62].
- Anti-quorum sensing activity is noticed in urolithins and ellagitannin metabolites [63].
- YenI/YenR system is observed in *Y. enterocolitica* [64].
- Virulence plasmid maintenance and hot cell attachment are also regulated by *Y. enterocolitica* quorum sensing process [65].

## 9.8 On *Yersinia pseudotuberculosis*

- YpsI/R and YtbI/R are LuxI/R homologues in *Yersinia pseudotuberculosis*. Quorum sensing regulates two temperature dependent phenomena such as motility and clumping [66].
- Quorum sensing regulates motility via YpsRI and YtbRI [67].
- Four T6SSs are identified in *Y. pseudotuberculosis*, which are thermoregulated. T6SS is regulated by quorum sensing, growth phase and temperature [68].
- Autoinducers of *Y. pseudotuberculosis* are analysed by a methodology called liquid chromatography—hybrid quadrupole linear ion trap [69].
- Experimental evidence shows that there is an interaction in quorum sensing, YmoA and LcrF. YmoA is a missing link, in quorum sensing regulated T3SS repression [70].

## 9.9 On *Aeromonas hydrophila*

- AhyI/R is a LuxI/R homologs in *Aeromonas hydrophila*, which controls gene expression [71].
- Exoprotease production is abolished by the autoinducers of *A. hydrophila* [72]
- *A. hydrophila* quorum sensing process regulates the biofilm development [73].
- Vanillin is a quorum sensing inhibitor, which reduces the *A. hydrophila* biofilm formation [74].
- Temperature controls the cell-to-cell communication process and biofilm formation of *A. hydrophila*. It has a very significant impact in food processing [75].

## 9.10 On *Aeromonas salmonicida*

- AsaI/R is a LuxI/R homologs in *Aeromonas salmonicida*, which controls gene expression [71].
- Quorum sensing regulates the virulence towards burbot (in *Aeromonas salmonicida* [76].
- Quorum sensing mechanism of *A. salmonicida* regulates pigment production and virulence in fish [77].
- Sulphur containing AHL-analogues is quorum sensing inhibitor for *A. salmonicida* [78].
- Spoilage and biofilm formation is regulated by *A. salmonicida* quorum sensing process [79].

## 9.11 On *Serratia liquefaciens*

- *Delisea pulchra* controls swarming motility and quorum sensing process of *Serratia liquefaciens* [80].
- Biofilm development is controlled by the cell-to cell communication process regulated genes *bsmB* and *bsmA* [81].
- *N*-hexanoyl-l-homoserine lactone and *N*-butanoyl-l-homoserine lactone are identified autoinducers in *Serratia liquefaciens* [82].
- *S. liquefaciens* biofilm formation is regulated by communication process, which is a cause of nosocomial infections. Large number of autoinducers production is not correlated with biofilm formations [83].
- Lemongrass oil cocktail with ampicillin is the most effect treatment (antibacterial activity) for *S. liquefaciens* [84].

## 9.12 On *Salmonella typhimurium*

- *S. typhimurium* produces signalling factor, which is used to two types of communication process [85].
- *S. typhimurium* are able to communicate with other bacterium using autoinducer-2 (interspecies communication) which are chemically distinct form of autoinducer [86].
- Malic and lactic acid is a quorum sensing inhibitors for interspecies communication process of *S. typhimurium* [87].
- Catalase expression is triggered by the quorum sensing process of *S. typhimurium* [88].
- Previous life cycle phase is significant for antibiotic resistance, growth kinetics, shape, cells death of *S. typhimurium* [89].

## 9.13 On *Ralstonia solanacearum*

- *R. solanacearum* is a phytopathogenic bacteria. The virulence gene expression is controlled by 3-hydroxypalmitic acid methyl ester [90].
- Aac is a quorum quenching agent for *R. solanacearum*, which has a possible application in biotechnology, agriculture [91].
- *phc* quorum sensing system is controlled by *(R)*-methyl 3-hydroxymyristate [92].
- 2-hydroxy-4-((methylamino)(phenyl)methyl) cyclopentanone is responsible for quorum sensing regulated biofilm formation in *R. solanacearum* [93].
- PhcA controls *R. solanacearum* Hrp genes [94].

## 9.14 On *Rhizobium etli*

- *R. etli* produces many autoinducers. 3OH-(slc)-HSL is one of the autoinducers [95].
- *R. etli* quorum sensing mechanism regulates symbiotic plasmid mobilization [96].
- The swarming behaviour of *R. etli* is regulated by autoinducers (as biosurfactants) [97].
- Several quorum sensing circuits are observed in *R. etli*, such as TraI/R, RaiI/R and CinI/R [98].
- Quorum sensing molecules of *R. etli* play a significant role in biofilm formation and motility [99].

## 9.15 On *Rhodobacter sphaeroides*

- CerI/R quorum sensing system is observed in photosynthetic bacteria *R. sphaeroides*, which is regulated by 7,8-cis-N-(tetradecenoyl)homoserine lactone [100].

## 9.16 On *Rhizobium leguminosarum*

- *cinRI* locus triggers the autoinducers production in *R. leguminosarum* [101]
- raiIR genes are part of the *R. leguminosarum* quorum sensing system [102].
- CinR, BisR and TraR are involved in plasmid transfer of *R. leguminosarum* [103].
- ExpR regulates *cin* and *rai* communication systems in *R. leguminosarum* [104].
- *R. leguminosarum* produces six different autoinducers [105].

## 9.17 On *Burkholderia cepacia*

- CepI/R is a LuxI/R homologs in pathogenic bacteria *B. cepacia* [106].
- cep quorum sensing process of *B. cepacia* controls swarming motility and biofilm formation [107].
- Communication process of *B. cepacia* controls the killing process of *C. elegans* [108].
- Diffusible signal factor plays a significant role in the communication process of *B. cepacia* [109].
- 4-hydroxy-3-methyl-2-alkylquinolines are produced by *B. cepacia* complex [110].

## 9.18 On *Chromobacterium violaceum*

- *C. violaceum* quorum sensing mechanism is regulated by HHL. HHL also regulates chitinolytic enzymes [111].
- LuxI/R type cell communication system is observed in *C. violaceum*, which is mediated by $C_{10}$-homoserine lactone. CviR receptor is identified which triggers gene expression [112].
- Carvacrol inhibits cell-to-cell bacterial communication in *C. violaceum* and truncates biofilm formation as well [113].
- Pigment extract from *A. auricular* inhibits *C. violaceum* quorum sensing process [114].
- Cell-to-cell communication process control of antimicrobial resistance [115].

## 9.19 On *Erwinia chrysanthemi*

- ExpI/R quorum sensing system is observed in phytopathogenic bacteria *E. chrysanthemi* [116].
- Autoinducer (AHL type signal) plays a significant role in *E. chrysanthemi* cell mortility, multicell aggregates and bacterial virulence [117].
- There autoinducers are identified in *E. chrysanthemi* [118].
- ExpR residues involve in the conformational changes and gives new insight of structure function relationship [119].
- *E. chrysanthemi* causes several plant diseases. The pathogenicity of *E. chrysanthemi* is triggered by pectinolytic enzymes [120].

## 9.20 On *Escherichia coli*

- Cross species communication is observed in *E. coli*. This type of communication is controlled by autoinducer 2. LuxS system is identified in *E. coli*. Furthermore, autoinducer 2 induces biofilm formation [121].

- A global regulatory communication mechanism is identified in *E. coli* with virulence factors [122].
- Quorum sensing molecules play an important role in *E. coli* mazEF mediated cell death [123].
- It has been reported that autoinducer 2 (with manipulation) affects on gut microbiota [124].
- QseBC communication system plays a role in the conversion of 3,4-Dihdroxymandelic acid from Norepinephrine [125].

## 9.21 On *Vibrio cholerae*

- *V. cholerae* is human pathogenous bacteria. ToxR signalling circuit regulates the virulence factors. LuxO and HapR regulate the communication mechanism [126].
- Three parallel quorum sensing systems exist in *V. cholerae* [127].
- CasA ,CsrB, CsrC and CsrD regulate the quorum sensing cascade in *V. cholerae* [128].
- Fis (small protein) takes part in the *V. cholerae* quorum sensing process [129].
- *V. cholerae* communication process also controls biofilm formation [130].

## 9.22 On *Streptococcus pneumonic*

- *S. pneumonic* is a gram-positive quorum sensing bacteria. The competence stimulating peptide is identified which induces *comCDE* expression [131].
- Competence specific genes and numerous quorum sensing output are linked uniquely by ComX [132].
- Two peptide pheromones are identified for intercellular communication which have different primary structures. ComABCDE and BlpABCSRH are two distinct signal pathways in *S. pneumonic*. These different communication systems regulate transcription of the same ABC transporter [133].
- *S. pneumoniae* is a human pathogen which is a cause of community acquired pneumonia. To inhibit the competence regulon in *S. pneumoniae* a new inhibitor (dnCSP) is designed [134].
- Quorum sensing is also regulated the bacteriocin secretion in *S. pneumonic* [135].

## 9.23 On *Bacillus subtilis*

- ComX is signalling molecules in *B. subtilis*. It stimulates competence. The structure of the ComX is reported [136].
- The important survival pathways are activated by competence and sporulation factor, quorum sensing pentapeptide of *B. subtilis* [137].

- Self-sensing is explored in a well quorum sensing systems of *B. subtilis* such as Rap-Phr and ComQXP. Here, bacterial cell directly responses to self-produced autoinducers in quorum independent and cell autonomous manner [138].
- ComQ protects against evolutionary competition. There is a link between response and signal in *B. subtilis* [139].
- A pump-probe signalling mechanism is identified in *B. subtilis* [140].

## 9.24 On *Staphylococcus aureus*

- *agr* quorum sensing system controls *S. aureus* biofilm formation which is a cause of bacterial infection diseases [141].
- Quorum sensing mechanism of *S. aureus* allows to sense confinement for survival [142].
- *S. aureus* quorum sensing process mediates the production of virulence factors. $\omega$- hydroxyemodin inhibits the quorum sensing process [143].
- *S. aureus* quorum sensing and biofilm development is controlled by surface attached molecules [144].
- Membrane embedded peptidase regulates the quorum sensing process of *S. aureus* [145].

## References

1. Schaefer AL, Hanzelka BL, Eberhard A, Greenberg EP (1996) Quorum sensing in *Vibrio fischeri*: probing autoinducer-LuxR interactions with autoinducer analogs. J Bacteriol 178(10):2897–2901
2. Schaefer AL, Val DL, Hanzelka BL, Cronan JE Jr, Greenberg EP (1996) Generation of cell-to-cell signals in quorum sensing: acyl homoserine lactone synthase activity of a purified *Vibrio fischeri* LuxI protein. Proc Natl Acad Sci. USA 93(18):9505
3. Lupp C, Ruby EG (2005) *Vibrio fischeri* uses two quorum-sensing systems for the regulation of early and late colonization factors. J Bacteriol 187(11):3620–3629
4. Hanzelka BL, Greenberg EP (1996) Quorum sensing in *Vibrio fischeri*: evidence that S-adenosylmethionines the amino acid substrate for autoinducer synthesis. J Bacteriol 178(17):5291–5294
5. Geske GD, O'Neill JC, Blackwell HE (2007) N-phenylacetanoyl-L-homoserine lactones can strongly antagonize or superagonize quorum sensing in *Vibrio fischeri*. ACS Chem Biol 2(5):315–319
6. Studer SV, Mandel MJ, Ruby EG (2008) AinS quorum sensing regulates the *Vibrio fischeri* acetate switch. J Bacteriol 190(17):5915–5923
7. Qin N, Callahan SM, Dunlap PV, Stevens AM (2007) Analysis of LuxR regulon gene expression during quorum sensing in *Vibrio fischeri*. J Bacteriol 189(11):4127–4134
8. Urbanowski ML, Lostroh CP, Greenberg EP (2004) Reversible acyl-homoserine lactone binding to purified *Vibrio fischeri* LuxR protein. J Bacteriol 186(3):631–637
9. Antunes LCM, Schaefer AL, Ferreira RB, Qin N, Stevens AM, Ruby EG, Greenberg EP (2007) Transcriptome analysis of the *Vibrio fischeri* LuxR-LuxI regulon. J Bacteriol 189(22):8387–8391

10. Castang S, Chantegrel B, Deshayes C, Dolmazon R, Gouet P, Haser R, Reverchon S, Nasser W, Hugouvieux-Cotte-Pattat N, Doutheau A (2004) N-Sulfonyl homoserine lactones as antagonists of bacterial quorum sensing. Bioorg Med Chem Lett 14(20):5145–5149
11. Pesci EC, Pearson JP, Seed PC, Iglewski BH (1997) Regulation of las and rhl quorum sensing in *Pseudomonas aeruginosa.* J Bacteriol 179(10):3127–3132
12. Whiteley M, Lee KM, Greenberg EP (1999) Identification of genes controlled by quorum sensing in *Pseudomonas aeruginosa.* Proc Natl Acad Sci 96(24):13904–13909
13. Pearson JP, Pesci EC, Iglewski BH (1997) Roles of *Pseudomonas aeruginosa* las and rhl quorum-sensing systems in control of elastase and rhamnolipid biosynthesis genes. J Bacteriol 179(18):5756–5767
14. Wagner VE, Bushnell D, Passador L, Brooks AI, Iglewski BH (2003) Microarray analysis of *Pseudomonas aeruginosa* quorum-sensing regulons: effects of growth phase and environment. J Bacteriol 185(7):2080–2095
15. Chugani SA, Whiteley M, Lee KM, D'Argenio D, Manoil C, Greenberg EP (2001) QscR, a modulator of quorum-sensing signal synthesis and virulence in *Pseudomonas aeruginosa.* Proc Natl Acad Sci 98(5):2752–2757
16. Williams P, Cámara M (2009) Quorum sensing and environmental adaptation in *Pseudomonas aeruginosa*: a tale of regulatory networks and multifunctional signal molecules. Curr Opin Microbiol 12(2):182–191
17. Singh PK, Schaefer AL, Parsek MR, Moninger TO, Welsh MJ, Greenberg EP (2000) Quorum-sensing signals indicate that cystic fibrosis lungs are infected with bacterial biofilms. Nature 407(6805):762
18. Tateda K, Comte R, Pechere JC, Köhler T, Yamaguchi K, Van Delden C (2001) Azithromycin inhibits quorum sensing in *Pseudomonas aeruginosa.* Antimicrob Agents Chemother 45(6):1930–1933
19. Dietrich LE, Price-Whelan A, Petersen A, Whiteley M, Newman DK (2006) The phenazine pyocyanin is a terminal signalling factor in the quorum sensing network of *Pseudomonas aeruginosa.* Mol Microbiol 61(5):1308–1321
20. Parsek MR, Greenberg EP (2005) Sociomicrobiology: the connections between quorum sensing and biofilms. Trends Microbiol 13(1):27–33
21. Schuster M, Hawkins AC, Harwood CS, Greenberg EP (2004) The *Pseudomonas aeruginosa* RpoS regulon and its relationship to quorum sensing. Mol Microbiol 51(4):973–985
22. Latifi A, Winson MK, Foglino M, Bycroft BW, Stewart GS, Lazdunski A, Williams P (1995) Multiple homologues of LuxR and LuxI control expression of virulence determinants and secondary metabolites through quorum sensing in *Pseudomonas aeruginosa* PAO1. Mol Microbiol 17(2):333–343
23. O'Loughlin CT, Miller LC, Siryaporn A, Drescher K, Semmelhack MF, Bassler BL (2013) A quorum-sensing inhibitor blocks *Pseudomonas aeruginosa* virulence and biofilm formation. Proc Natl Acad Sci 110(44):17981–17986
24. Chen R, Déziel E, Groleau MC, Schaefer AL, Greenberg EP (2019) Social cheating in a *Pseudomonas aeruginosa* quorum-sensing variant. Proc Natl Acad Sci 116(14):7021–7026
25. Wang M, Schaefer AL, Dandekar AA, Greenberg EP (2015) Quorum sensing and policing of *Pseudomonas aeruginosa* social cheaters. Proc Natl Acad Sci 112(7):2187–2191
26. Bodelón G, Montes-García V, López-Puente V, Hill EH, Hamon C, Sanz-Ortiz MN, Rodal-Cedeira S, Costas C, Celiksoy S, Pérez-Juste I, Scarabelli L (2016) Detection and imaging of quorum sensing in *Pseudomonas aeruginosa* biofilm communities by surface-enhanced resonance Raman scattering. Nat Mater 15(11):1203
27. Maura D, Hazan R, Kitao T, Ballok AE, Rahme LG (2016) Evidence for direct control of virulence and defense gene circuits by the *Pseudomonas aeruginosa* quorum sensing regulator, MvfR. Sci Rep 6:34083
28. Rampioni G, Falcone M, Heeb S, Frangipani E, Fletcher MP, Dubern JF, Visca P, Leoni L, Cámara M, Williams P (2016) Unravelling the genome-wide contributions of specific 2-alkyl-4-quinolones and PqsE to quorum sensing in *Pseudomonas aeruginosa.* PLoS Pathog 12(11):e1006029

29. Singh BR, Singh BN, Singh A, Khan W, Naqvi AH, Singh HB (2015) Mycofabricated biosilver nanoparticles interrupt *Pseudomonas aeruginosa* quorum sensing systems. Sci Rep 5:13719
30. Darch SE, Simoska O, Fitzpatrick M, Barraza JP, Stevenson KJ, Bonnecaze RT, Shear JB, Whiteley M (2018) Spatial determinants of quorum signaling in a *Pseudomonas aeruginosa* infection model. Proc Natl Acad Sci 115(18):4779–4784
31. Zhang HB, Wang LH, Zhang LH (2002) Genetic control of quorum-sensing signal turnover in *Agrobacterium tumefaciens*. Proc Natl Acad Sci 99(7):4638–4643
32. Chevrot R, Rosen R, Haudecoeur E, Cirou A, Shelp BJ, Ron E, Faure D (2006) GABA controls the level of quorum-sensing signal in *Agrobacterium tumefaciens*. Proc Natl Acad Sci 103(19):7460–7464
33. An D, Danhorn T, Fuqua C, Parsek MR (2006) Quorum sensing and motility mediate interactions between Pseudomonas aeruginosa and *Agrobacterium tumefaciens* in biofilm cocultures. Proc Natl Acad Sci 103(10):3828–3833
34. Zhu J, Winans SC (1998) Activity of the quorum-sensing regulator TraR of *Agrobacterium tumefaciens* is inhibited by a truncated, dominant defective TraR-like protein. Mol Microbiol 27(2):289–297
35. Zhu J, Chai Y, Zhong Z, Li S, Winans SC (2003) Agrobacterium bioassay strain for ultra-sensitive detection of N-acylhomoserine lactone-type quorum-sensing molecules: detection of autoinducers in *Mesorhizobium huakuii*. Appl Environ Microbiol 69(11):6949–6953
36. Liu D, Thomas PW, Momb J, Hoang QQ, Petsko GA, Ringe D, Fast W (2007) Structure and specificity of a quorum-quenching lactonase (AiiB) from *Agrobacterium tumefaciens*. Biochemistry 46(42):11789–11799
37. El Sahili A, Li SZ, Lang J, Virus C, Planamente S, Ahmar M, Guimaraes BG, Aumont-Nicaise M, Vigouroux A, Soulère L, Reader J (2015) A pyranose-2-phosphate motif is responsible for both antibiotic import and quorum-sensing regulation in *Agrobacterium tumefaciens*. PLoS Pathog 11(8):e1005071
38. Dessaux Y, Faure D (2018) Quorum sensing and quorum quenching in *Agrobacterium*: a "go/no go system"? Genes 9(4):210
39. Zhang B, Guo B, Bai Y, Lu H, Dong Y (2018) Synthesis and biological evaluation of azamacrolide comprising the triazole moiety as quorum sensing inhibitors. Molecules 23(5):1086
40. McCarthy RR, Yu M, Eilers K, Wang YC, Lai EM, Filloux A (2019) Cyclic di-GMP inactivates T6SS and T4SS activity in *Agrobacterium tumefaciens*. Mol Microbiol 112(2):632–648
41. Byers JT, Lucas C, Salmond GP, Welch M (2002) Nonenzymatic turnover of an *Erwinia carotovora* quorum-sensing signaling molecule. J Bacteriol 184(4):1163–1171
42. Burr T, Barnard AM, Corbett MJ, Pemberton CL, Simpson NJ, Salmond GP (2006) Identification of the central quorum sensing regulator of virulence in the enteric phytopathogen, *Erwinia carotovora*: the VirR repressor. Mol Microbiol 59(1):113–125
43. Sjöblom S, Brader G, Koch G, Palva ET (2006) Cooperation of two distinct ExpR regulators controls quorum sensing specificity and virulence in the plant pathogen *Erwinia carotovora*. Mol Microbiol 60(6):1474–1489
44. Zhang Y, Kong J, Xie Y, Guo Y, Cheng Y, Qian H, Yao W (2018) Essential oil components inhibit biofilm formation in *Erwinia carotovora* and Pseudomonas fluorescens via anti-quorum sensing activity. LWT 92:133–139
45. Zhang Y, Kong J, Xie Y, Guo Y, Yu H, Cheng Y, Qian H, Shi R, Yao W (2019) Quorum-sensing inhibition by hexanal in biofilms formed by *Erwinia carotovora* and *Pseudomonas fluorescens*. LWT 109:145–152
46. Surette MG, Miller MB, Bassler BL (1999) Quorum sensing in *Escherichia coli*, *Salmonella typhimurium*, and *Vibrio harveyi*: a new family of genes responsible for autoinducer production. Proc Natl Acad Sci 96(4):1639–1644
47. Henke JM, Bassler BL (2004) Three parallel quorum-sensing systems regulate gene expression in *Vibrio harveyi*. J Bacteriol 186(20):6902–6914

48. Mok KC, Wingreen NS, Bassler BL (2003) *Vibrio harveyi* quorum sensing: a coincidence detector for two autoinducers controls gene expression. EMBO J 22(4):870–881
49. Henke JM, Bassler BL (2004) Quorum sensing regulates type III secretion in *Vibrio harveyi* and *Vibrio parahaemolyticus*. J Bacteriol 186(12):3794–3805
50. Lilley BN, Bassler BL (2000) Regulation of quorum sensing in *Vibrio harveyi* by LuxO and sigma-54. Mol Microbiol 36(4):940–954
51. Waters CM, Bassler BL (2006) The Vibrio harveyi quorum-sensing system uses shared regulatory components to discriminate between multiple autoinducers. Genes Dev 20(19):2754–2767
52. van Kessel JC, Rutherford ST, Cong JP, Quinodoz S, Healy J, Bassler BL (2015) Quorum sensing regulates the osmotic stress response in *Vibrio harveyi*. J Bacteriol 197(1):73–80
53. Bagert JD, van Kessel JC, Sweredoski MJ, Feng L, Hess S, Bassler BL, Tirrell DA (2016) Time-resolved proteomic analysis of quorum sensing in *Vibrio harveyi*. Chem Sci 7(3):1797–1806
54. McRose DL, Baars O, Seyedsayamdost MR, Morel FM (2018) Quorum sensing and iron regulate a two-for-one siderophore gene cluster in *Vibrio harveyi*. Proc Natl Acad Sci 115(29):7581–7586
55. Salini R, Santhakumari S, Ravi AV, Pandian SK (2019) Synergistic antibiofilm efficacy of undecanoic acid and auxins against quorum sensing mediated biofilm formation of luminescent *Vibrio harveyi*. Aquaculture 498:162–170
56. Milton DL, Hardman A, Camara M, Chhabra SR, Bycroft BW, Stewart GS, Williams P (1997) Quorum sensing in Vibrio anguillarum: characterization of the vanI/vanR locus and identification of the autoinducer N-(3-oxodecanoyl)-L-homoserine lactone. J Bacteriol 179(9):3004–3012
57. Weber B, Hasic M, Chen C, Wai SN, Milton DL (2009) Type VI secretion modulates quorum sensing and stress response in *Vibrio anguillarum*. Environ Microbiol 11(12):3018–3028
58. Weber B, Croxatto A, Chen C, Milton DL (2008). RpoS induces expression of the *Vibrio anguillarum* quorum-sensing regulator VanT. Microbiology 154(3):767–780
59. Croxatto A, Pride J, Hardman A, Williams P, Cámara M, Milton DL (2004) A distinctive dual-channel quorum-sensing system operates in *Vibrio anguillarum*. Mol Microbiol 52(6):1677–1689
60. Tan D, Svenningsen SL, Middelboe M (2015) Quorum sensing determines the choice of antiphage defense strategy in *Vibrio anguillarum*. MBio 6(3):e00627-15
61. Atkinson S, Chang CY, Sockett RE, Cámara M, Williams P (2006) Quorum sensing in *Yersinia enterocolitica* controls swimming and swarming motility. J Bacteriol 188(4):1451–1461
62. Truchado P, Giménez Bastida JA, Larrosa M, Castro-Ibáñez I, Espín JC, Tomás Barberán FA, García-Conesa MT, Allende A (2012) Inhibition of quorum sensing (QS) in *Yersinia enterocolitica* by an orange extract rich in glycosylated flavanones. J Agric Food Chem 60(36):8885–8894
63. Giménez-Bastida JA, Truchado P, Larrosa M, Espín JC, Tomás-Barberán FA, Allende A, García-Conesa MT (2012) Urolithins, ellagitannin metabolites produced by colon microbiota, inhibit quorum sensing in *Yersinia enterocolitica*: phenotypic response and associated molecular changes. Food Chem 132(3):1465–1474
64. Throup JP, Camara M, Briggs GS, Winson MK, Chhabra SR, Bycroft BW, Williams P, Stewart GS (1995) Characterisation of the yenI/yenR locus from *Yersinia enterocolitica* mediating the synthesis of two N-acylhomoserine lactone signal molecules. Mol Microbiol 17(2):345–356
65. Ng YK, Grasso M, Wright V, Garcia V, Williams P, Atkinson S (2018) The quorum sensing system of *Yersinia enterocolitica* 8081 regulates swimming motility, host cell attachment, and virulence plasmid maintenance. Genes 9(6):307
66. Atkinson S, Throup JP, Stewart GS, Williams P (1999) A hierarchical quorum-sensing system in *Yersinia pseudotuberculosis* is involved in the regulation of motility and clumping. Mol Microbiol 33(6):1267–1277

67. Atkinson S, Chang CY, Patrick HL, Buckley CM, Wang Y, Sockett RE, Cámara M, Williams P (2008) Functional interplay between the *Yersinia pseudotuberculosis* YpsRI and YtbRI quorum sensing systems modulates swimming motility by controlling expression of flhDC and fliA. Mol Microbiol 69(1):137–151
68. Zhang W, Xu S, Li J, Shen X, Wang Y, Yuan Z (2011) Modulation of a thermoregulated type VI secretion system by AHL-dependent quorum sensing in *Yersinia pseudotuberculosis*. Arch Microbiol 193(5):351–363
69. Ortori CA, Atkinson S, Chhabra SR, Cámara M, Williams P, Barrett DA (2007) Comprehensive profiling of N-acylhomoserine lactones produced by *Yersinia pseudotuberculosis* using liquid chromatography coupled to hybrid quadrupole–linear ion trap mass spectrometry. Anal. Bioanal. Chem 387(2):497–511
70. Slater A (2018) Unraveling the regulatory relationship between quorum sensing and the type III secretion system in *Yersinia pseudotuberculosis* (Doctoral dissertation, University of Nottingham)
71. Swift S, Karlyshev AV, Fish L, Durant EL, Winson MK, Chhabra SR, Williams P, Macintyre S, Stewart GS (1997) Quorum sensing in *Aeromonas hydrophila* and *Aeromonas salmonicida*: identification of the LuxRI homologs AhyRI and AsaRI and their cognate N-acylhomoserine lactone signal molecules. J Bacteriol 179(17):5271–5281
72. Swift S, Lynch MJ, Fish L, Kirke DF, Tomás JM, Stewart GS, Williams P (1999) Quorum sensing-dependent regulation and blockade of exoprotease production in *Aeromonas hydrophila*. Infect Immun 67(10):5192–5199
73. Lynch MJ, Swift S, Kirke DF, Keevil CW, Dodd CE, Williams P (2002) The regulation of biofilm development by quorum sensing in *Aeromonas hydrophila*. Environ Microbiol 4(1):18–28
74. Ponnusamy K, Paul D, Kweon JH (2009) Inhibition of quorum sensing mechanism and *Aeromonas hydrophila* biofilm formation by vanillin. Environ Eng Sci 26(8):1359–1363
75. Mizan MFR, Jahid IK, Park SY, Silva JL, Kim TJ, Myoung J, Ha SD (2018) Effects of temperature on biofilm formation and quorum sensing of *Aeromonas hydrophila*. Ital J Food Sci 30:456–466
76. Natrah FMI, Alam MI, Pawar S, Harzevili AS, Nevejan N, Boon N, Sorgeloos P, Bossier P, Defoirdt T (2012) The impact of quorum sensing on the virulence of *Aeromonas hydrophila* and *Aeromonas salmonicida* towards burbot (*Lota lota* L.) larvae. Vet Microbiol 159(1–2):77–82
77. Schwenteit J, Gram L, Nielsen KF, Fridjonsson OH, Bornscheuer UT, Givskov M, Gudmundsdottir BK (2011) Quorum sensing in *Aeromonas salmonicida* subsp. achromogenes and the effect of the autoinducer synthase AsaI on bacterial virulence. Vet Microbiol 147(3–4):389–397
78. Rasch M, Kastbjerg VG, Bruhn JB, Dalsgaard I, Givskov M, Gram L (2007) Quorum sensing signals are produced by *Aeromonas salmonicida* and quorum sensing inhibitors can reduce production of a potential virulence factor. Dis Aquat Org 78(2):105–113
79. Liu L, Yan Y, Feng L, Zhu J (2018) Quorum sensing asaI mutants affect spoilage phenotypes, motility, and biofilm formation in a marine fish isolate of *Aeromonas salmonicida*. Food Microbiol 76:40–51
80. Rasmussen TB, Manefield M, Andersen JB, Eberl L, Anthoni U, Christophersen C, Steinberg P, Kjelleberg S, Givskov M (2000). How Delisea pulchra furanones affect quorum sensing and swarming motility in *Serratia liquefaciens* MG1. Microbiology 146(12):3237–3244
81. Labbate M, Queck SY, Koh KS, Rice SA, Givskov M, Kjelleberg S (2004) Quorum sensing-controlled biofilm development in *Serratia liquefaciens* MG1. J Bacteriol 186(3):692–698
82. Eberl L, Winson MK, Sternberg C, Stewart GS, Christiansen G, Chhabra SR, Bycroft B, Williams P, Molin S, Givskov M (1996) Involvement of N-acyl-l-homoserine lactone autoinducers in controlling the multicellular behaviour of *Serratia liquefaciens*. Mol Microbiol 20(1):127–136

83. Remuzgo-Martínez, S, Lázaro-Díez M, Mayer C, Aranzamendi-Zaldumbide M, Padilla D, Calvo J, Marco F, Martínez-Martínez L, Icardo JM, Otero A, Ramos-Vivas J (2015) Biofilm formation and quorum-sensing-molecule production by clinical isolates of *Serratia liquefaciens*. Appl Environ Microbiol 81(10):3306–3315
84. Araz K, Patel S, Khajehdehi N, Bjerkan Wade C, Thomas J, Krajnyk I (2018) Interactions between lemongrass and lavender essential oils in combination with ampicillin influencing antibacterial activity on *Sporosarcina ureae* and *Serratia liquefaciens*. WURJ Health Nat Sci 9(1):2
85. Surette MG, Bassler BL (1998) Quorum sensing in *Escherichia coli* and *Salmonella typhimurium*. Proc Natl Acad Sci 95(12):7046–7050
86. Miller ST, Xavier KB, Campagna SR, Taga ME, Semmelhack MF, Bassler BL, Hughson FM (2004) *Salmonella typhimurium* recognizes a chemically distinct form of the bacterial quorum-sensing signal AI-2. Mol Cell 15(5):677–687
87. Almasoud A, Hettiarachchy N, Rayaprolu S, Babu D, Kwon YM, Mauromoustakos A (2016) Inhibitory effects of lactic and malic organic acids on autoinducer type 2 (AI-2) quorum sensing of *Escherichia coli* O157: H7 and *Salmonella typhimurium*. LWT-Food Sci Technol 66:560–56
88. Liao H, Zhong X, Xu L, Ma Q, Wang Y, Cai Y, Guo X (2019) Quorum-sensing systems trigger catalase expression to reverse the oxyR deletion-mediated VBNC state in *Salmonella typhimurium*. Res Microbiol 170(2):65–73
89. Hawkins JL, Uknalis J, Oscar TP, Schwarz JG, Vimini B, Parveen S (2019) The effect of previous life cycle phase on the growth kinetics, morphology, and antibiotic resistance of *Salmonella typhimurium* DT104 in brain heart infusion and ground chicken extract. Front Microbiol 10:1043
90. Flavier AB, Clough SJ, Schell MA, Denny TP (1997) Identification of 3-hydroxypalmitic acid methyl ester as a novel autoregulator controlling virulence in *Ralstonia solanacearum*. Mol Microbiol 26(2):251–259
91. Chen CN, Chen CJ, Liao CT, Lee CY (2009) A probable aculeacin A acylase from the *Ralstonia solanacearum* GMI1000 is N-acyl-homoserine lactone acylase with quorum-quenching activity. BMC Microbiol 9(1):89
92. Kai K, Ohnishi H, Shimatani M, Ishikawa S, Mori Y, Kiba A, Ohnishi K, Tabuchi M, Hikichi Y (2015) Methyl 3-hydroxymyristate, a diffusible signal mediating phc quorum sensing in *Ralstonia solanacearum*. ChemBioChem 16(16):2309–2318
93. Kumar JS, Umesha S, Prasad KS, Niranjana P (2016) Detection of quorum sensing molecules and biofilm formation in *Ralstonia solanacearum*. Curr Microbiol 72(3):297–305
94. Genin S, Brito B, Denny TP, Boucher C (2005) Control of the *Ralstonia solanacearum* Type III secretion system (Hrp) genes by the global virulence regulator PhcA. FEBS Lett 579(10):2077–2081
95. Daniels R, De Vos DE, Desair J, Raedschelders G, Luyten E, Rosemeyer V, Verreth C, Schoeters E, Vanderleyden J, Michiels J (2002) The cin quorum sensing locus of *Rhizobium etli* CNPAF512 affects growth and symbiotic nitrogen fixation. J Biol Chem 277(1):462–468
96. Tun-Garrido C, Bustos P, Gonzáäez V, Brom S (2003) Conjugative transfer of p42a from *Rhizobium etli* CFN42, which is required for mobilization of the symbiotic plasmid, is regulated by quorum sensing. J Bacteriol 185(5):1681–1692
97. Daniels R, Reynaert S, Hoekstra H, Verreth C, Janssens J, Braeken K, Fauvart M, Beullens S, Heusdens C, Lambrichts I, De Vos DE (2006) Quorum signal molecules as biosurfactants affecting swarming in *Rhizobium etli*. Proc Natl Acad Sci 103(40):14965–14970
98. Zheng H, Mao Y, Zhu Q, Ling J, Zhang N, Naseer N, Zhong Z, Zhu J (2015) The quorum sensing regulator CinR hierarchically regulates two other quorum sensing pathways in ligand-dependent and-independent fashions in *Rhizobium etli*. J Bacteriol 197(9):1573–1581
99. Dixit S, Dubey RC, Maheshwari DK, Seth PK, Bajpai VK (2017) Roles of quorum sensing molecules from *Rhizobium etli* RT1 in bacterial motility and biofilm formation. Brazil J Microbiol 48(4):815–821

100. Puskas A, Greenberg EÁ, Kaplan S, Schaefer AÁ (1997) A quorum-sensing system in the free-living photosynthetic bacterium *Rhodobacter sphaeroides*. J Bacteriol 179(23):7530–7537
101. Lithgow JK, Wilkinson A, Hardman A, Rodelas B, Wisniewski-Dyé F, Williams P, Downie JA (2000) The regulatory locus cinRI in *Rhizobium leguminosarum* controls a network of quorum-sensing loci. Mol Microbiol 37(1):81–97
102. Wisniewski-Dye F, Jones J, Chhabra SR, Downie JA (2002) railR genes are part of a quorum-sensing network controlled by cinI and cinR in *Rhizobium leguminosarum*. J Bacteriol 184(6):1597–1606
103. McAnulla C, Edwards A, Sanchez-Contreras M, Sawers RG, Downie JA (2007) Quorum-sensing-regulated transcriptional initiation of plasmid transfer and replication genes in *Rhizobium leguminosarum* biovar viciae. Microbiology 153(7):2074–2082
104. Edwards A, Frederix M, Wisniewski-Dyé F, Jones J, Zorreguieta A, Downie JA (2009) The cin and rai quorum-sensing regulatory systems in *Rhizobium leguminosarum* are coordinated by ExpR and CinS, a small regulatory protein coexpressed with CinI. J Bacteriol 191(9):3059–3067
105. Blosser-Middleton RS, Gray KM (2001) Multiple N-acyl homoserine lactone signals of *Rhizobium leguminosarum* are synthesized in a distinct temporal pattern. J Bacteriol 183(23):6771–6777
106. Lewenza S, Conway B, Greenberg EP, Sokol PA (1999) Quorum sensing in *Burkholderia cepacia*: identification of the LuxRI homologs CepRI. J Bacteriol 181(3):748–756
107. Huber B, Riedel K, Hentzer M, Heydorn A, Gotschlich A, Givskov M, Molin S, Eberl L (2001) The cep quorum-sensing system of *Burkholderia cepacia* H111 controls biofilm formation and swarming motility. Microbiology 147(9):2517–2528
108. Köthe M, Antl M, Huber B, Stoecker K, Ebrecht D, Steinmetz I, Eberl L (2003) Killing of Caenorhabditis elegans by *Burkholderia cepacia* is controlled by the cep quorum-sensing system. Cellular Microbiol 5(5):343–351
109. Deng Y, Wu JE, Eberl L, Zhang LH (2010) Structural and functional characterization of diffusible signal factor family quorum-sensing signals produced by members of the *Burkholderia cepacia* complex. Appl Environ Microbiol 76(14):4675–4683
110. Coulon PM, Déziel E (2019) Potential of the *Burkholderia cepacia* complex to produce 4-hydroxy-3-methyl-2-alkyquinolines. Front Cell Infect Microbiol 9:33
111. Chernin LS, Winson MK, Thompson JM, Haran S, Bycroft BW, Chet I,Williams P, Stewart GS (1998) Chitinolytic activity in *Chromobacterium violaceum*: substrate analysis and regulation by quorum sensing. J Bacteriol 180(17):4435–4441
112. Stauff DL, Bassler BL (2011) Quorum sensing in *Chromobacterium violaceum*: DNA recognition and gene regulation by the CviR receptor. J Bacteriol 193(15):3871–3878
113. Burt SA, Ojo-Fakunle VT, Woertman J, Veldhuizen EJ (2014) The natural antimicrobial carvacrol inhibits quorum sensing in *Chromobacterium violaceum* and reduces bacterial biofilm formation at sub-lethal concentrations. PLoS One 9(4):e93414
114. Zhu H, He CC, Chu QH (2011) Inhibition of quorum sensing in *Chromobacterium violaceum* by pigments extracted from *Auricularia auricular*. Lett Appl Microbiol 52(3):269–274
115. Evans KC, Benomar S, Camuy-Vélez LA, Nasseri EB, Wang X, Neuenswander B, Chandler JR (2018) Quorum-sensing control of antibiotic resistance stabilizes cooperation in *Chromobacterium violaceum*. ISME J 12(5):1263
116. Reverchon S, Bouillant ML, Salmond G, Nasser W (1998) Integration of the quorum-sensing system in the regulatory networks controlling virulence factor synthesis in *Erwinia chrysanthemi*. Mol Microbiol 29(6):1407–1418
117. Hussain MB, Zhang HB, Xu JL, Liu Q, Jiang Z, Zhang LH (2008) The acyl-homoserine lactone-type quorum-sensing system modulates cell motility and virulence of *Erwinia chrysanthemi* pv. zeae. J Bacteriol 190(3):1045–1053

118. Nasser W, Bouillant ML, Salmond G, Reverchon S (1998) Characterization of the *Erwinia chrysanthemi* expI–expR locus directing the synthesis of two N-acyl-homoserine lactone signal molecules. Mol Microbiol 29(6):1391–1405
119. Castang S, Reverchon S, Gouet P, Nasser W (2006) Direct evidence for the modulation of the activity of the *Erwinia chrysanthemi* quorum-sensing regulator ExpR by acylhomoserine lactone pheromone. J Biol Chem 281(40):29972–29987
120. Hugouvieux-Cotte-Pattat N, Condemine G, Nasser W, Reverchon S (1996) Regulation of pectinolysis in *Erwinia chrysanthemi*. Annu Rev Microbiol 50(1):213–257
121. Barrios AFG, Zuo R, Hashimoto Y, Yang L, Bentley WE, Wood TK (2006) Autoinducer 2 controls biofilm formation in *Escherichia coli* through a novel motility quorum-sensing regulator (MqsR, B3022). J Bacteriol 188(1):305–316
122. Sperandio V, Torres AG, Girón JA, Kaper JB (2001) Quorum sensing is a global regulatory mechanism in enterohemorrhagic *Escherichia coli* O157: H7. J Bacteriol 183(17):5187–5197
123. Kolodkin-Gal I, Hazan R, Gaathon A, Carmeli S, Engelberg-Kulka H (2007) A linear pentapeptide is a quorum-sensing factor required for mazEF-mediated cell death in *Escherichia coli*. Science 318(5850):652–655
124. Thompson JA, Oliveira RA, Djukovic A, Ubeda C, Xavier KB (2015) Manipulation of the quorum sensing signal AI-2 affects the antibiotic-treated gut microbiota. Cell Rep 10(11):1861–1871
125. Pasupuleti S, Sule N, Manson MD, Jayaraman A (2018) Conversion of norepinephrine to 3, 4-dihdroxymandelic acid in *Escherichia coli* requires the QseBC quorum-sensing system and the fear transcription factor. J Bacteriol 200(1):e00564-17
126. Zhu J, Miller MB, Vance RE, Dziejman M, Bassler BL, Mekalanos JJ (2002) Quorum-sensing regulators control virulence gene expression in *Vibrio cholerae*. Proc Natl Acad Sci 99(5):3129–3134
127. Miller MB, Skorupski K, Lenz DH, Taylor RK, Bassler BL (2002) Parallel quorum sensing systems converge to regulate virulence in *Vibrio cholerae*. Cell 110(3):303–314
128. Lenz DH, Miller MB, Zhu J, Kulkarni RV, Bassler BL (2005) CsrA and three redundant small RNAs regulate quorum sensing in *Vibrio cholerae*. Mol Microbiol 58(4):1186–1202
129. Lenz DH, Bassler BL (2007) The small nucleoid protein Fis is involved in *Vibrio cholerae* quorum sensing. Mol Microbiol 63(3):859–871
130. Hammer BK, Bassler BL (2003) Quorum sensing controls biofilm formation in *Vibrio cholerae*. Mol Microbiol 50(1):101–104
131. Alloing G, Martin B, Granadel C, Claverys JP (1998) Development of competence in *Streptococcus pneumoniae*: pheromone autoinduction and control of quorum sensing by the oligopeptide permease. Mol Microbiol 29(1):75–83
132. Luo P, Li H, Morrison DA (2003) ComX is a unique link between multiple quorum sensing outputs and competence in *Streptococcus pneumoniae*. Mol Microbiol 50(2):623–633
133. Knutsen E, Ween O, Håvarstein LS (2004) Two separate quorum-sensing systems upregulate transcription of the same ABC transporter in *Streptococcus pneumoniae*. J Bacteriol 186(10):3078–3085
134. Yang Y, Lin J, Harrington A, Cornilescu G, Lau GW, Tal-Gan Y (2020) Designing cyclic competence-stimulating peptide (CSP) analogs with pan-group quorum-sensing inhibition activity in *Streptococcus pneumoniae*. Proc Natl Acad Sci 117(3):1689–1699
135. Miller EL, Kjos M, Abrudan MI, Roberts IS, Veening JW, Rozen DE (2018) Eavesdropping and crosstalk between secreted quorum sensing peptide signals that regulate bacteriocin production in *Streptococcus pneumoniae*. ISME J 12(10):2363–2375
136. Okada M, Sato I, Cho SJ, Iwata H, Nishio T, Dubnau D, Sakagami Y (2005) Structure of the *Bacillus subtilis* quorum-sensing peptide pheromone ComX. Nat Chem Biol 1(1):23–24
137. Fujiya M, Musch MW, Nakagawa Y, Hu S, Alverdy J, Kohgo Y, Schneewind O, Jabri B, Chang EB (2007) The Bacillus subtilis quorum-sensing molecule CSF contributes to intestinal homeostasis via OCTN2, a host cell membrane transporter. Cell Host Microbe 1(4):299–308

138. Bareia T, Pollak S, Eldar A (2018) Self-sensing in Bacillus subtilis quorum-sensing systems. Nat Microbiol 3(1):83–89
139. Oslizlo A, Stefanic P, Dogsa I, Mandic-Mulec I (2014) Private link between signal and response in Bacillus subtilis quorum sensing. Proc Natl Acad Sci 111(4):1586–1591
140. Babel H, Naranjo-Meneses P, Trauth S, Schulmeister S, Malengo G, Sourjik V, Bischofs IB (2020) Ratiometric population sensing by a pump-probe signaling system in *Bacillus subtilis*. Nat Commun 11(1):1–13
141. Yarwood JM, Bartels DJ, Volper EM, Greenberg EP (2004) Quorum sensing in *Staphylococcus aureus* biofilms. J Bacteriol 186(6):1838–1850
142. Carnes EC, Lopez DM, Donegan NP, Cheung A, Gresham H, Timmins GS, Brinker CJ (2010) Confinement-induced quorum sensing of individual *Staphylococcus aureus* bacteria. Nat Chem Biol 6(1):41–45
143. Daly SM, Elmore BO, Kavanaugh JS, Triplett KD, Figueroa M, Raja HA, El-Elimat T, Crosby HA, Femling JK, Cech NB, Horswill AR (2015) $\omega$-Hydroxyemodin limits Staphylococcus aureus quorum sensing-mediated pathogenesis and inflammation. Antimicrob Agents Chemother 59(4):2223–2235
144. Kim MK, Zhao A, Wang A, Brown ZZ, Muir TW, Stone HA, Bassler BL (2017) Surface-attached molecules control *Staphylococcus aureus* quorum sensing and biofilm development. Nat Microbiol 2(8):17080
145. Cosgriff CJ, White CR, Teoh WP, Grayczyk JP, Alonzo F (2019) Control of *Staphylococcus aureus* quorum sensing by a membrane-embedded peptidase. Infection Immunity 87(5):e00019-19

# Therapy Related Mathematical Models and Quorum Quenching

# 10

**Abstract**

Bacterial communication systems regulate virulence factors and play a key role in different bacterial infection diseases. In this chapter, we address the concept of quorum quenching and discuss mathematical models of anti-quorum sensing which are useful for the treatment purpose. Models predictions are significant for the future experimentations.

## 10.1 Quorum Quenching

Bacteria use quorum sensing mechanism to achieve maximal benefit in this highly competitive environment. We have already discussed in details about the bacterial optimal survival strategies, which we called quorum sensing. Many types of autoinducers (or signal) are already been discovered. AHL is one of them. AHLs control and coordinate the communication systems. This bacterial conversation process (or quorum sensing) regulates several other bacterial behaviour (i.e. biofilms formation, virulence). To take an upper hand in this competition biologists discover quorum sensing inhibitors and quorum quenching enzymes. These inhibitors and/ or enzymes are the key elements about the concept of quorum quenching. The concept of quorum quenching has a great impact to control bacterial infections. The first quorum quenching enzyme is discovered in the year 2000. The enzyme is encoded by *aiia* gene [1]. After that several quorum quenching enzymes are identified. These enzymes have a significant role in pathogen–host and microbe–microbe interactions [2, 3]. The research in the context of AHL-degradation enzymes can shed new light to overcome several bacterial infection diseases [2, 3]. In future, we have to be more focussed on the regulation and the role of the quorum quenching enzymes in their host and study the system under natural condition.


S. Majumdar, S. Roy, *Microbial Communication*,
https://doi.org/10.1007/978-981-15-7417-7_10

## 10.2 Mathematical Framework of Anti-quorum Sensing

We have already discussed several mathematical formalism of bacterial quorum sensing mechanism with other regulated phenomena in the past chapters in the book. Mathematical modellers have specific aim to formulate a model based on assumptions which are scientifically relevant and motivate mechanisms of the systems. The investigation is running with equation and/or system of equations to keep on eye whether it can regenerate experimental observations and shed new light to uncover the systems. We address the anti-quorum sensing model in batch culture, biofilm development (early stage) and mature biofilm development [4–6]. The mathematical derivation is based on the *P. aeruginosa* LasI/R system which is useful for LuxI/R systems. The total bacterial population is divided into two subdivision one is up-regulated cells and the other one is down-regulated cell [7]. We summarise the assumptions as follows:

- Down-regulated cells are represented by $\tilde{N}_d$. Bacterial cell contains empty *lux*-box. Autoinducers and EPS matrix is produced by *P. aeruginosa* at low rate. In this situation, we do not observe any virulence activity.
- Up-regulated cells are represented by $\tilde{N}_u$. Bacterial cells have complex (Autoinducer-LasR) bound lux-box. Autoinducers and EPS are generated at enhanced rate. We also observe a virulent activity.
- $\tilde{N}_T$ is the total bacterial population and $\tilde{N}_T = \tilde{N}_d + \tilde{N}_u$.
- Down-regulated cell divided into two down-regulated cells.
- Up-regulated cell divided into one up-regulated and one down-regulated cell.
- $\tilde{R}$ represents the concentration of LasR which is produced at the rate $\tilde{R}_0$. It binds with autoinducer $\tilde{A}$ and form a complex $\tilde{P}$ ( Autoinducer-LasR). We have $\frac{d\tilde{R}}{dt} = \tilde{R}_0 - \tilde{k}_{ra}\tilde{A}\tilde{R} + \tilde{k}_p\tilde{P} - \tilde{\lambda}_{\tilde{R}}\tilde{R}$ and complex (Autoinducer-LasR) equation $\frac{d\tilde{P}}{dt} = \tilde{k}_{ra}\tilde{A}\tilde{R} - \tilde{k}_p\tilde{P} - \tilde{\lambda}_p\tilde{P}$. Output of LasI happens at constant rate and decay is given by $\frac{d\tilde{L}}{dt} = \tilde{L}_0 - \tilde{\lambda}_{\tilde{L}}\tilde{L}$.
- Autoinducers are produced with $\tilde{k}_d$ a background level and decay with a constant $\tilde{\lambda}$. The rate of change of autoinducer (down-regulated) is expressed as $\tilde{k}_d - \tilde{k}_{ra}\tilde{A}\tilde{R} + \tilde{k}_p\tilde{P} - \tilde{\lambda}\tilde{A}$.
- The rate of change of autoinducers (up-regulated cells) is expressed as $\tilde{k}_a\tilde{L} + \tilde{k}_d - \tilde{k}_{ra}\tilde{A}\tilde{R} + \tilde{k}_p\tilde{P} - \tilde{\lambda}\tilde{A}$ where $\tilde{k}_a\tilde{L}$ represents a massive increase (up-regulated cells) of autoinducers generation. The rate of change of autoinducers (external media) is given by $-\tilde{\lambda}\tilde{A}$.
- If $\frac{d\tilde{R}}{dt} = \frac{d\tilde{P}}{dt} = \frac{d\tilde{L}}{dt} = 0$ then we get $\tilde{L} = \tilde{L}_\infty$, $\tilde{P} = \frac{\tilde{P}_\infty}{\tilde{R}_\infty}\tilde{R}\tilde{A}$, $\tilde{R} = \frac{\tilde{R}_\infty}{1+\tilde{\mu}_{\tilde{R}}\tilde{A}}$ where $\tilde{L}_\infty = \frac{\tilde{L}_0}{\tilde{\lambda}_{\tilde{L}}}$, $\tilde{R}_\infty = \frac{\tilde{R}_0}{\tilde{\lambda}_{\tilde{R}}}$, $\tilde{\mu}_{\tilde{R}} = \frac{\tilde{\lambda}_p\tilde{P}_\infty}{\tilde{\lambda}_{\tilde{R}}\tilde{R}_\infty}$ and $\tilde{P}_\infty = \frac{\tilde{R}_\infty\tilde{k}_{ra}}{(\tilde{k}_p+\tilde{\lambda}_p)}$.
- Substitute $\tilde{R} = \tilde{R}_\infty$, $\tilde{\mu}_{\tilde{R}}\tilde{A} \approx 0$, $\tilde{P} = \tilde{P}_\infty\tilde{A}$, we get the rate of change of autoinducers (down-regulated cells) is expressed as $\tilde{k}_d - \tilde{\sigma}\tilde{A} - \tilde{\lambda}\tilde{A}$ and the rate

of change of autoinducers (up-regulated) is expressed as $\tilde{k}_u + \tilde{k}_d - \tilde{\sigma}\tilde{A} - \tilde{\lambda}\tilde{A}$, where $\tilde{\sigma} = \tilde{\lambda}_p \tilde{P}_\infty$ and $\tilde{k}_u = \tilde{k}_a \tilde{L}_\infty$.

- We consider that up-regulation rate of cells is proportional to $\tilde{P}_\infty \tilde{A}$ (complex concentration). Thus, $\tilde{\alpha}\tilde{A}$ is the up-regulation rate where $\tilde{\alpha} = \tilde{\alpha}_a \tilde{P}_\infty$. $\tilde{\alpha}_a$ is the proportional constant. $\tilde{\beta} = \tilde{\lambda}_p$ is the down-regulation rate of cell.
- Assume $\tilde{Q}_1$, $\tilde{Q}_2$, $\tilde{Q}_3$ be the concentration of the anti-LuxR ( homolog) agent, anti-autoinducer agent and anti-LuxI ( homolog) agent, respectively, which follows

$$\text{LasR} + \tilde{Q}_1 \xrightarrow{\tilde{k}_1} (1 - \tilde{v}_1)\tilde{Q}_1 + \text{ by product}$$
$$\text{Autoinducer} + \tilde{Q}_2 \xrightarrow{\tilde{\mu}_2} (1 - \tilde{v}_2)\tilde{Q}_2 + \text{ by product}$$
$$\text{Autoinducer} + \tilde{Q}_3 \xrightarrow{\tilde{\mu}_3} (1 - \tilde{v}_3)\tilde{Q}_3 + \text{ by product}$$

where $\tilde{v}_1$, $\tilde{v}_2$ and $\tilde{v}_3$ denote the average amount of $\tilde{Q}_1$, $\tilde{Q}_2$, $\tilde{Q}_3$ lost by reaction, respectively.

- LasR concentration is given by $\tilde{R} = \frac{\tilde{R}_\infty}{(1+\tilde{\gamma}_1\tilde{Q}_1)}$ at equilibrium, where $\tilde{\gamma}_1 = \frac{\tilde{k}_1}{\tilde{k}_{\tilde{R}}}$. Autoinducers-LasR binding rate and up-regulation rate are reduced by a factor $(1 + \tilde{\gamma}_1\tilde{Q}_1)$.
- The equation of rate of change of autoinducers has an additional term $-\tilde{\mu}_2\tilde{Q}_2\tilde{A}$.
- LasI equilibrium level reduces to $\frac{\tilde{L}_\infty}{(1+\tilde{\gamma}_3\tilde{Q}_3)}$, where $\tilde{\gamma}_3 = \frac{\tilde{k}_3}{\tilde{\lambda}_L}$. Thus $\frac{\tilde{k}_u}{(1+\tilde{\gamma}_3\tilde{Q}_3)}$ is the new autoinducer output rate term.

Now, the governing mathematical model of anti-quorum sensing treatment in batch culture is expressed as

$$\frac{d\tilde{N}_d}{dt} = \tilde{r}\tilde{N}_T - \frac{\tilde{\alpha}\tilde{A}}{1+\tilde{\gamma}_1\tilde{Q}_1}\tilde{N}_d + \tilde{\beta}\tilde{N}_u \tag{10.1}$$

$$\frac{d\tilde{N}_u}{dt} = \frac{\tilde{\alpha}\tilde{A}}{1+\tilde{\gamma}_1\tilde{Q}_1}\tilde{N}_d - \tilde{\beta}\tilde{N}_u \tag{10.2}$$

$$\frac{d\tilde{A}}{dt} = \frac{\tilde{k}_u}{1+\tilde{\gamma}_3\tilde{Q}_3}\tilde{N}_u + \tilde{k}_d\tilde{N}_T - \frac{\tilde{\sigma}\tilde{A}}{1+\tilde{\gamma}_1\tilde{Q}_1}\tilde{N}_T - \tilde{\lambda}\tilde{A} - \tilde{\mu}_2\tilde{Q}_2\tilde{A} \tag{10.3}$$

$$\frac{d\tilde{Q}_1}{dt} = \tilde{\phi}_1 - \frac{\tilde{\mu}_1\tilde{Q}_1}{1+\tilde{\gamma}_1\tilde{Q}_1}\tilde{N}_T - \tilde{\lambda}_1\tilde{Q}_1 \tag{10.4}$$

$$\frac{d\tilde{Q}_2}{dt} = \tilde{\phi}_2 - \tilde{\mu}_2\tilde{v}_2\tilde{A}\tilde{Q}_2 - \tilde{\lambda}_2\tilde{Q}_2 \tag{10.5}$$

$$\frac{d\tilde{Q}_3}{dt} = \tilde{\phi}_3 - \frac{\tilde{\mu}_3\tilde{Q}_3}{1+\tilde{\gamma}_3\tilde{Q}_3}\tilde{N}_u - \tilde{\lambda}_3\tilde{Q}_3 \tag{10.6}$$

This mathematical model is continuous in space-time. We neglect the stochastic effects. Drug is introduced at the beginning or begin drip-fed at a rate $\tilde{\phi}_i$ (for $i = 1, 2, 3$). The drug loss rates are denoted by the parameters $\tilde{\mu}_1 = \tilde{v}_1\tilde{k}_1$ and $\tilde{\mu}_3 = \tilde{v}_3\tilde{k}_3$.

Let us consider the bacterial cell distribution as a function of $z$ (space) and $t$ (time). $z$ gives us a perpendicular distance from the bacterial biofilm base with $z = H(t)$. The death cells occupy a volume fraction ($\tilde{M}$). The rest of the space is filled by EPS ($\tilde{E}$) and water ($\tilde{W}$). Hence, $\tilde{N}_T + \tilde{M} + \tilde{E} + \tilde{W} = 1$. The pore space is increasing at the time of EPS production. This way , we have $\tilde{W} = \tilde{W}_0 + \tilde{\Theta}\tilde{E}$ where $\tilde{W}_0$ and $\tilde{\Theta}$ are constant. Finally, we get $\tilde{N}_T + \tilde{M} + (1 + \tilde{\Theta})\tilde{E} = 1 - \tilde{W}_0$. Moreover, assume that bacterial quorum sensing mechanism control the nutrient concentration ($\tilde{c}$) and EPS production [7]. The mathematical formalism of anti-quorum sensing treatment in biofilm is given by

$$\frac{\partial \tilde{N}_T}{\partial t} + \frac{\partial \tilde{v}\tilde{N}_T}{\partial z} = \tilde{N}_T(\tilde{F}_b(\tilde{c}) - \tilde{F}_d(\tilde{c})) \tag{10.7}$$

$$\frac{\partial \tilde{M}}{\partial t} + \frac{\partial \tilde{v}\tilde{M}}{\partial z} = \tilde{N}_T\tilde{F}_d(\tilde{c}) \tag{10.8}$$

$$\frac{\partial \tilde{N}_u}{\partial t} + \frac{\partial \tilde{v}\tilde{N}_u}{\partial z} = \frac{\tilde{\alpha}\tilde{A}}{1 + \tilde{\gamma}_1\tilde{Q}_1}\tilde{N}_d - \tilde{\beta}\tilde{N}_u \tag{10.9}$$

$$\frac{\partial \tilde{E}}{\partial t} + \frac{\partial \tilde{v}\tilde{E}}{\partial z} = (\tilde{E}_0\tilde{N}_T + \tilde{k}_{\tilde{E}}\tilde{N}_u)\tilde{F}_b(\tilde{c}) - \tilde{\lambda}_{\tilde{E}}\tilde{E} \tag{10.10}$$

$$0 = \tilde{D}_a\frac{\partial^2 \tilde{A}}{\partial z^2} + \frac{\tilde{k}_u^*}{1 + \tilde{\gamma}_3\tilde{Q}_3}\tilde{N}_u + \tilde{k}_d^*\tilde{N}_T - \frac{\tilde{\sigma}^*\tilde{A}}{1 + \tilde{\gamma}_1\tilde{Q}_1}\tilde{N}_T - \tilde{\lambda}\tilde{A} - \tilde{\mu}_2\tilde{Q}_2\tilde{A} \tag{10.11}$$

$$0 = \tilde{D}_1\frac{\partial^2 \tilde{Q}_1}{\partial z^2} - \frac{\tilde{\mu}_1^*\tilde{Q}_1}{1 + \tilde{\gamma}_1\tilde{Q}_1}\tilde{N}_T - \tilde{\lambda}_1\tilde{Q}_1 \tag{10.12}$$

$$0 = \tilde{D}_3\frac{\partial^2 \tilde{Q}_3}{\partial z^2} - \frac{\tilde{\mu}_3^*\tilde{Q}_3}{1 + \tilde{\gamma}_3\tilde{Q}_3}\tilde{N}_u - \tilde{\lambda}_3\tilde{Q}_3 \tag{10.13}$$

$$0 = \tilde{D}_{\tilde{c}}\frac{\partial^2 \tilde{c}}{\partial z^2} - \tilde{\rho}\tilde{N}_T\tilde{F}_b(\tilde{c}) \tag{10.14}$$

$$\frac{\partial \tilde{v}}{\partial z} = \frac{1}{1 - \tilde{W}_0}\left(\tilde{N}_T\tilde{F}_b(\tilde{c}) + (1 + \tilde{\Theta})(\tilde{E}_0\tilde{N}_T + \tilde{k}_{\tilde{E}}\tilde{N}_u)\tilde{F}_b(\tilde{c}) - \tilde{\lambda}_{\tilde{E}}\tilde{E}\right) \tag{10.15}$$

$$\frac{dH}{dt} = \tilde{v}(H, t) \tag{10.16}$$

**Table 10.1** List of parameters use in the model

| Variables | Descriptions |
|---|---|
| $\tilde{\rho}$ | Oxygen consumption constant |
| $\tilde{\tau}$ | Sets minimal death rate |
| $\tilde{c}_1$ | Birth rate oxygen concentration (Half max.) |
| $\tilde{c}_2$ | Death rate oxygen concentration (Half max.) |
| $\tilde{c}_{ext}$ | Dissolved oxygen concentration |
| $\tilde{B}_1$ | Maximum birth rate |
| $\tilde{B}_2$ | Maximum death rate |
| $\tilde{Q}_a$ | Surface autoinducer mass transfer rate |
| $\tilde{D}_a$ | Diffusion rate of autoinducer |
| $\tilde{D}_i$ | Diffusion rate of species |
| $\tilde{D}_{\tilde{c}}$ | Diffusion rate of oxygen |
| $\tilde{E}_0$ | Background EPS production rate |
| $\tilde{k}_{\tilde{E}}$ | Max. EPS production rate |
| $\tilde{\lambda}_{\tilde{E}}$ | EPS decay rate |
| $\tilde{\omega}$ | Maximum density of cells in biofilms |
| $\tilde{\Theta}$ | EPS generated pore space constant |
| $\tilde{H}_0$ | Initial biofilm depth |
| $\tilde{\lambda}_i$ | Decay rate of quorum sensing inhibitor |
| $\tilde{\varphi}_i$ | Drip rate of quorum sensing inhibitor |
| $\tilde{\upsilon}_1$ | Mean $Q_2$ loss in reaction with autoinducer |
| $\tilde{\gamma}_1 \cdot \tilde{\gamma}_3$ | 1/conc. when quorum sensing inhibitor is 50% effective |
| $\tilde{\mu}_1, \tilde{\mu}_2, \tilde{\mu}_3$ | Drug loss rates (due to quorum sensing inhibitor action) |
| $\tilde{\sigma}$ | Autoinducer loss rate by Autoinducer/LasR binding |
| $\tilde{\lambda}$ | Autoinducer decay rate |
| $\tilde{k}_d$ | Autoinducer production rate by up-regulated cells |
| $\tilde{\beta}$ | Down-regulation rate |
| $\tilde{\alpha}$ | Maximal up-regulation rate |
| $\tilde{r}$ | Cell birth rate |

We analyse the mathematical framework with Michaelis–Menten kinetics $\tilde{F}_d(\tilde{c}) = \tilde{B}_2(1 - \tilde{\tau}\frac{\tilde{c}}{\tilde{c}_2+\tilde{c}})$, $\tilde{F}_b(\tilde{c}) = \tilde{B}_1\frac{\tilde{c}}{\tilde{c}_1+\tilde{c}}$ where $\tilde{F}_d(\tilde{c})$ and $\tilde{F}_b(\tilde{c})$ denote the death and birth rate, respectively (Table 10.1).

Finally, several predictions of the mathematical formalism are listed as

- After the initial period, up-regulation begins. It is observed that up-regulation is rapidly started after 3 h and 12–13% up-regulation is noticed at any time. Moreover, up-regulation depends on the growth phase in batch culture [7].
- Bacterial colony virulence is calculated by
  $\tilde{N}_u^{frac} = 1 - \frac{\tilde{\sigma}(\tilde{\beta}+\tilde{r})}{\tilde{\alpha}\tilde{k}_u}$ (for exponential growth phase)
  $\tilde{N}_u^{frac} = 1 - \frac{\tilde{\beta}(\tilde{\sigma}\tilde{K}+\tilde{\lambda})}{\tilde{\alpha}\tilde{k}_u\tilde{K}}$ (for stationary phase) ($\tilde{K}$ is the population size) [7].

- Anti-LasI agent is more effective than others. Anti-LasR treatment is the most effective quorum sensing inhibitor (QSI).
- Bacterial biofilm growth is slowed down after initial rise up. Up-regulated cell fraction is expressed as $\tilde{U}(t) = \frac{\int_0^H \tilde{N}_u(z,t)dz}{\int_0^H \tilde{N}_T(z,t)dz}$. Living cells are located near surface.
- A shift is observed in biofilm growth rates with scale $\mu M$. Anti-LasR and Anti-LasI agent are similar [7].
- QSIs suppress the bacterial communication (or quorum sensing) in batch culture as well as in biofilms.

## References

1. Dong, Y. H., Xu, J. L., Li, X. Z., & Zhang, L. H. (2000). AiiA, an enzyme that inactivates the acylhomoserine lactone quorum-sensing signal and attenuates the virulence of Erwinia carotovora. Proceedings of the National Academy of Sciences, 97(7), 3526–3531.
2. Dong, Y. H., & Zhang, L. H. (2005). Quorum sensing and quorum-quenching enzymes. The Journal of Microbiology, 43(1), 101–109.
3. Dong, Y. H., Wang, L. H., & Zhang, L. H. (2007). Quorum-quenching microbial infections: mechanisms and implications. Philosophical transactions of the Royal Society B: biological Sciences, 362(1483), 1201–1211.
4. Anguige, K., King, J. R., Ward, J. P., & Williams, P. (2004). Mathematical modelling of therapies targeted at bacterial quorum sensing. Mathematical biosciences, 192(1), 39–83.
5. Anguige, K., King, J. R., & Ward, J. P. (2005). Modelling antibiotic-and anti-quorum sensing treatment of a spatially-structured Pseudomonas aeruginosa population. Journal of mathematical biology, 51(5), 557–594.
6. Anguige, K., King, J. R., & Ward, J. P. (2006). A multi-phase mathematical model of quorum sensing in a maturing Pseudomonas aeruginosa biofilm. Mathematical biosciences, 203(2), 240–276.
7. Ward, J. (2008). Mathematical modeling of quorum-sensing control in biofilms. In Control of biofilm infections by signal manipulation (pp. 79–108). Springer, Berlin, Heidelberg.

# Role of Noise in Microbial Communication 11

**Abstract**

Bacterial communication systems are very interactive which control several biological functions. Noise is an inherent property of biological communication systems. This chapter deals with the effect of noise in the bacterial communication mechanism with the help of weakly non-local hydrodynamics and viscosity. The theoretical framework is based on the relationship among the living fluid, non-local noise and the noisy Burgers equation.

## 11.1 Introduction

Historically, bacteria have been considered primarily as autonomous unicellular organisms with limited collective behavioural ability. Presently it has become clear that bacteria are, in fact, highly interactive. The term quorum sensing has been adopted to describe the bacterial communication process which coordinates, among other variables, gene expression. Such event is usually, but not always expressed, when the population reached a high cell density. Quorum sensing bacteria produce and release chemical signal molecules, referred to as autoinducers, that increase in concentration as a function of cell density. The detection of a threshold autoinducer concentration leads to an alteration in gene expression. These processes include symbiosis, virulence, competence, conjugation antibiotic production, motility, sporulation and biofilm formation.

Observational results indicate that swimming bacteria in suspensions develop a coordinated motion. In other words, water fluidity is modified, in a nontrivial manner, by the presence of bacteria, above a threshold number density. At such threshold condition swimming bacterial suspensions impose a coordinated water movement on a length scale of the order of (10–100) $\mu$m compared with a bacterial size of the order of 3 $\mu$m [1]. This observation leads to fundamental questions concerning the cell-to-cell communication mechanism among bacteria,


S. Majumdar, S. Roy, *Microbial Communication*,
https://doi.org/10.1007/978-981-15-7417-7_11

presently known as quorum sensing. Bassler and her group [2, 3] regard the role of the chemicals released by bacteria (known as quorum sensing molecules) as evoking behavioural modification on its neighbours. Their study centres on Vibrio harveyi (which is a gram-negative, bioluminescent, marine bacterium) as a possible approach to analyse the mechanism of multi-signal integration. Recently, Alberghini et al. [4] critically reviewed the current observations and have considered the various assumptions behind the theoretical framework for quorum sensing.

The early observations of Vibrio fisheri's (a species of bioluminescent bacterium) behaviour where bacterial growth within a host fish body lead to a concept of minimum cell density. In this hypothesis such density must attain critical threshold value before all subsequent activities take place. This kind of interpretation is usually based on two assumptions:

- The presence of a homogenous system such that signal density be close to isotropic, allowing informational simultaneity along the entire population.
- Growth should be congruent with the density increase, i.e. the environment of the suspended bacteria must be such that cells do not actively migrate or disperse.

However, the study of substrate geometry indicates that the above paradigm may not be optimal to address bacterial communication in biofilm. And so, an alternative interpretation of "diffusion sensing" has been proposed [5]. Alberghini et al. [4] studied the consequences of critical cellular distribution and calculated the gradient profiles, based on the mechanism of diffusion of autoinducer molecules, and concluded diffusion sensing is a viable hypothesis. Hydrodynamic model of "swimming" bacteria or bacterial colonies [1, 6]seems to be one of the most comprehensive alternative models defining possible quorum sensing mechanism. Here densely packed bacteria may be viewed as a "bacterial fluid" or "living fluid" similar to that of dense granular systems. Lega and Passot [6] initially assumed a two-phase hydrodynamic equations taken the bacteria and water as two interpenetrating and interacting continuum. However, by considering the relatively high bacterial density, given the fact that no water motion is observed (under isothermal conditions) in absence of the bacteria, we assume the dynamics of the suspended bacteria is governed by bacterial dynamics. Under these conditions bacteria and water appear to move as a single fluid at hydrodynamic scale. Moreover, since the bacteria are of finite size this fluid is considered as granular medium. However, the behaviour of the fluid of granular mass is different from that of typical fluid. Indeed, the actual granularity imposes a second source of fluctuations other than thermal fluctuations; as grains cannot be treated as points, at any length scale. The finite size of the grains (here bacteria) is the source of this fluctuation. This type of fluctuation is known as non-local noise in contrast to local noise in usual hydrodynamic flow. So we have two type of noise—one due to thermal fluctuations and the other the non-thermal one due to existence of finite size of the bacteria. These two types of noise are always present and on the prevalence of any one depends on the nature of the force applied to the system. The usual hydrodynamical equations need to be modified so as to accommodate this non-local noise. It is worth mentioning that some investigations have already been done [7] to study the behaviour of this type of hydrodynamics usually known as non-local hydrodynamics. In this chapter we will focus on the role of this particular type of noise, i.e. non-thermal noise in microbial communication within the framework of non-local hydrodynamics. At first we discuss Weakly Non-local Hydrodynamics and collective behaviour of bacteria for convenience in the next section.

## 11.2 Weakly Non-local Hydrodynamics and Collective Behaviour of Bacteria

Closed packed populations of suspended swimming bacteria develop a coordinated motion on length scales (10–100) μm in comparison to the size of an individual bacterium of the order of 3 μm when the concentration of the bacteria reaches a sufficiently high value. Recently, Koch and Subramanian [1] discussed the collective hydrodynamics of such living organisms called *living fluids* and tried to understand its behaviour based on numerical analysis and stability analysis. We emphasize that the behaviour of the closed packed bacteria is similar to the behaviour of dense granular systems. It is well known that dense granular systems behave like a fluid in a manner fundamentally different from that of ordinary fluid. The existence of intermediate length scale due to the fact that the grains cannot be considered as points at any length scale lead us to introduce second source of fluctuations in addition to thermodynamic fluctuations.

### 11.2.1 Non-local Noise

This second type of fluctuation is known as non-local noise and has been discussed extensively in the context of flow of granular materials. Both thermal and non-thermal noise are present in the system and the dominance of one on the other depends on the forcing $F_g = \frac{f}{\rho g}$ applied to the system where $f$ be the volume density of the forcing and $g$ the acceleration due to gravity. Hydrodynamical equations have been studied for the both the limits i.e. strong and weak forcing limits. In case of closely packed bacteria though the size of the bacterium is smaller than that of the system but it cannot be considered as point. The existence of intermediate scale associated with the finite size of the bacteria gives rise to non-local noise. Moreover, the swimming induced stresses of the bacteria may change the local arrangement of bacteria and induce stress fluctuations. These stress fluctuations may lead to shear motion somewhere else far away from the source and hence called non-local. This type of non-locality is different from the non-locality in quantum theory. The rearrangement occurring due to shear stress is considered as self-activated process. The shear due to this self-activated process makes gene expression possible. The above analysis leads us to formulate a non-local hydrodynamic framework to study the collective behaviour of bacteria so as to understand quorum sensing. Weakly non-local extension of hydrodynamics has been studied by several authors in the context of continuum mechanics with internal structures, coarse graining or gradient of the density and its higher order derivatives. This type of noise is shown to be related to the kinematic viscosity of the living fluid. It plays an important role in understanding the cell-to-cell communication in a comprehensive manner. Now let us discuss the weakly non-local hydrodynamics and viscosity.

### 11.2.2 Weakly Non-local Hydrodynamics and Viscosity

The state space of one component fluid is described by the density $\rho$ and $\mathbf{v}$, the velocity of fluid. The state space $(\rho, \mathbf{v})$ and velocity gradient span the domain of constitutive functions. The stress tensor or the pressure term is the only constitutive quantity in this framework. The weakly non-local extension of this framework is done by considering the higher order derivatives of the basic variables like density and velocity in constitutive equations. In fluid dynamical framework the balance of mass and balance of momentum are usually expressed as

$$\dot{\rho} + \rho \nabla \dot{\mathbf{v}} = \sigma_m \tag{11.1}$$

and

$$\rho \dot{\mathbf{v}} + \nabla \dot{P} = \rho \mathbf{f} \tag{11.2}$$

respectively. Here $P$ is the pressure and $\mathbf{f}$ be the force density. This is known as Cauchy momentum equation. Now we extend this framework by considering the state space spanned by the variables

$$(\rho, \nabla \rho, \mathbf{v}, \nabla \mathbf{v}, \nabla^2 \rho).$$

The balance of mass and balance of momentum can be rewritten taking the time and space derivatives of the constitutive variables denoted by

$$(\dot{\rho}, \nabla \dot{\rho}, \dot{\mathbf{v}}, \nabla \dot{\mathbf{v}}, \nabla^2 \mathbf{v}, \nabla^3 \rho)$$

It has been proved by Van and Fulop [10] there exists a scalar valued function $\phi_v$ or non-local potential such that

$$\nabla \cdot \sigma = -\nabla \phi_v \tag{11.3}$$

where $\phi_v$ is the course-grained potential or kinematic viscosity and $\sigma_{ij}$ be the shear tensor. This viscosity can be calculated from the entropy function as:

$$\phi_v = \nabla.(\rho \partial \nabla \rho s) - \partial_\rho (\rho s) \tag{11.4}$$

where $s$ is the entropy. Choosing particular form of $s$ we can study specific non-local fluids. We can choose the simplest potential function using the following entropy density function

$$s(\rho, \nabla\rho) = -\nu\frac{\nabla\rho^2}{4\rho} - \frac{\mathbf{v}^2}{2} \tag{11.5}$$

The non-local potential can be deduced from this entropy as

$$-\frac{\nu}{2}\nabla^2\rho \tag{11.6}$$

$\nu$ is a constant coefficient. Let us define a *kinematic velocity* $\mathbf{u_k}$ as

$$\mathbf{u_k} = \nabla\rho$$

The kinematic velocity is introduced to relate to a kind of fluctuations (non-thermal) due to the course-grained nature of the fluid. After some algebraic calculations we can write a general expression

$$\nabla\mathbf{u} + (\mathbf{u}.\nabla)\mathbf{u} = \nu\nabla^2\mathbf{u} + \nabla\eta \tag{11.7}$$

where

$$\nabla\eta = -\nu\nabla^2(\Delta\mathbf{u_k}).$$

Here,

$$\Delta\mathbf{u_k} = \mathbf{u} - \mathbf{u_k}$$

The above equation is noisy Burgers equation. Now we will discuss this framework to understand phenomena of quorum sensing among the bacteria.

## 11.3 Quorum Sensing and Metastable State

The bacteria confined in a particular volume can change their environment when they reach a particular density. Experimental observations clearly indicate the existence of some cooperative activities of the cells if they reach to a particular density. This cooperation can change the environment and produces coordinated action when sufficient density of cells is available. Here, the cell-to-cell communication is used for this coordinated action. Bacteria release some chemical molecules known as signalling molecules or more precisely autoinducers. The term quorum sensing was coined by Fuqua et al. [8] to indicate the minimum behaviourial unit of cells as a quorum of bacteria. There exists various *apparently* conflicting hypothesis regarding the function of this type of signalling. Quorum Sensing (QS) and Diffusion Sensing (DS) are generally considered as two competing hyoptheses. Burkhard et al. [9] proposed a new explanation unifying these two "apparently conflicting hypotheses" called Efficiency Sensing (ES). Recently, West et al. [10] critically analysed the three proposals, i.e. QS, DS and ES and claimed that ES is not required to unite QS and DS. We emphasize that the framework of non-local hydrodynamics can explain the phenomena from a more physical perspective in a constant way. It is evident from our analysis in previous section that the rearrangement of the configuration of the coarse grained systems produce a noise which gives rise to kinematic viscosity.

This kind of viscosity has been first introduced to understand disk like structure formation at galactic scale by Zeldovich [11] and popularly known as adhesive approximation. Saluena et al. [12] studied this type of noise in connection with the hydrodynamics of dense granular systems. We emphasize that this type of noise plays a crucial role in understanding quorum sensing of the cells in the following manner.

Closely packed cells or bacteria in suspension interact through their hydrodynamic disturbance when the concentration becomes sufficiently high. The colony of bacteria seems to interact with the medium say water in which they are suspended and one needs to take care of this interaction in understanding the behaviour. Since no motion is observed in the medium in the absence of the bacteria or if the bacteria are dead and also due to the fact that the density of bacteria is high, it is reasonable to assume that the bacteria and the water moves as a single fluid. However, the behaviour of this fluid is different from ideal fluid. This is similar to the behaviour of fluid associated with the dense granular systems since there exists a non-local (non-thermal) noise due finite size of the bacterium and hence an intermediate scale. In non-local hydrodynamics for dense granular materials the state space is considered as function of $(\rho, \nabla\rho, \mathbf{v})$ and other higher order derivatives of density and velocity. The non-locality is mathematically expressed through these higher order derivatives. It is to be noted that the kinematic viscosity as a gradient function in non-local hydrodynamics. Now to understand the significance of this more realistically within the domain of bacterial communication [13] we need to study position based sensing of the bacteria.

### 11.3.1 Position Based Sensing

The discovery of autoinducer molecules laid down the physical basis of cell-to-cell bacterial communication. The diffusion of these autoinducers have been studied by several groups where it is possible to calculate the gradient profiles either around a single cell or at the centre of increasing densities of cell. Alberghini et al. [4] studied positional sensing of the cells vs. quorum sensing and cell-to-cell communication. They emphasized that one should consider cell-to-cell communication in terms of positional sensing where instead of considering concentration of a mixed bulk solution it is necessary to take care of the topologically distinct values sensed by individual cell at different positions of the environment. This type of gradient topology plays an important role in cell-to-cell communication. In the framework of non-local hydrodynamics, one considers the gradient of density as another state variable for the fluid description of densely packed bacteria. We define a velocity similar to kinematic velocity in the framework of non-local hydrodynamics as

$$\mathbf{u_{kin}} = \nabla \rho$$

where $\rho$ is the density of the fluid. Because of existence of non-local noise and non-thermal noise (arising from the intermediate scale due to finite size of the bacterium) $\mathbf{u_{kin}}$ velocity will vary and

$$\Delta \mathbf{u_k} \equiv \mathbf{u} - \mathbf{u_k}$$

where $\mathbf{u}$ is the mean velocity. This will give rise to a viscosity

$$\nabla \eta \equiv -\nu \nabla^2 (\Delta \mathbf{u_k}).$$

This viscosity forms a metastable state of the bacteria similar to adhesive approximation of Zeldovich in producing a disk at galactic scale. Of course, the two scales are different, i.e. galactic scale is much bigger than the scale pertaining to the metastable state of bacteria. Here, the noise term will add an extra term in Burgers equation and the stability of noisy Burgers equation has been studied with small viscosity[14].

### 11.3.2 Metastable State and Threshold Density

The multi-time scale behaviours of Burgers equation with small viscosity by Beck et al. [14] indicates the existence of metastable states of physical system. The stationary or stable states of physical systems are well studied in applied mathematics. The stability of the states occur for the viscosity

$$0.01 < \nu < 0.1.$$

Again this kinematic viscosity is the ratio of dynamic viscosity and the density. If we put the value of dynamic viscosity for particular medium, the density can be estimated for quorum sensing state. Moreover, the dynamic viscosity depends on the temperature. For various temperatures of the medium, threshold density will also vary so as to maintain fixed value of viscosity. However, if we consider particular density, we get stability of state even at various temperatures. This is very important from the point of view of understanding neuronal activities of brain. Long and Fee [15] analysed the temporal dynamics of motor pathway of song bird controlling the temperature and tried to find the neural circuitry that controls the behaviours at different time scales. This may open new vistas in understanding brain function.

The metastable states are not fixed points of the equation of motion but a family of states which emerge relatively quickly and ultimately goes to asymptotic state. Here, multi-time scales are important in the sense that at one time scale metastable state emerges and at another a family of such states and for another asymptotic states emerge. In two dimensional fluid flows the metastable states play important role. In case of bacteria the kinematic viscosity (as the ratio of dynamic viscosity and the density) will be small since the density is very high to form quorum. The emergence of metastable state is possible for this kind of physical system with high density (small viscosity) for short time scale. During this time period as the states are metastable, the fluctuation of stress due to autoinducing molecules will produce a fluctuation in the configuration of the system which induce shear somewhere else. This happens in non-local rheology. This process is a self-activated process.

### 11.3.3 Self-activated Process and Ratio of Bulk to Shear Viscosity

The shear force exerted by the autoinducing molecules will produce fluctuation of the stress in the configuration of the densely packed bacteria. As soon as the concentration of the autoinducers reach a threshold it leads to the change of the arrangement of the configuration and leads to change in bulk viscosity. The bulk viscosity for dense fluid is related to irreversible resistance. This is different from thermal fluctuation since thermal fluctuation is negligible for dense fluid. This change in bulk viscosity will produce shear to somewhere else. In densely granular medium the fluctuation of the stress and consequent production of shear is considered to be a self-activated process. In the present scenario of densely packed bacteria the ratio of bulk viscosity and the shear viscosity is associated with this type of self-activated process which makes the expression of genes possible and synchronization happens[13]. Iberall and Llinas[16] postulated that the basic organizing force for the living system controls dynamically the ratio of bulk to shear viscosity.

It is to be noted that gene regulation due to shear stress has become an active area of research. More than 40 genes have been identified to be modulated by the shear stress. In the context of quorum sensing the genes have been already identified which express as soon as density of the cells reach a threshold value. The expression of these genes in QS can be tested in laboratory experiments by changing shear stress.

## 11.4 Possible Implications

It is evident from the above analysis that we need to consider two major steps in understanding quorum sensing of bacteria. As the density of the bacteria becomes high and reaches a threshold, a metastable state is formed for a particular time period and then the ratio of bulk to shear viscosity associated with a self-regulatory process due to stress fluctuation by autoinducer molecules gives rise to shear making gene expression possible. The metastable state is produced for small viscosity arising out of the internal non-local noise. Here we have single framework where one need not to consider QS, DS or ES separately for the description of cell-to-cell bacteria communication. It is worth mentioning that the adhesive approximation which is used in the context of disk formation at galactic scale is also useful in understanding the metastable state for cell-to-cell communication. It is to be noted that non-local noise is associated with kinematic viscosity and for certain range of this kinematic viscosity quorum of the bacteria occurs. It sheds new light on the issues of applicability of physical principles in biological systems.

## References

1. Koch DL, Subramanian G (2011) Collective hydrodynamics of swimming microorganisms: living fluids. Ann Rev Fluid Mech 43:637–659
2. Waters CM, Bassler BL (2005) Quorum sensing: cell-to-cell communication in bacteria. Ann Rev Cell Dev Biol 21:319–346
3. Long T, Tu KC, Wang Y, Mehta P, Ong NP, Bassler BL, Wingreen NS (2009) Quantifying the integration of quorum-sensing signals with single-cell resolution. PLoS Biol 7(3):e1000068
4. Alberghini S, Polone E, Corich V, Carlot M, Seno F, Trovato A, Squartini A (2009) Consequences of relative cellular positioning on quorum sensing and bacterial cell-to-cell communication. FEMS Microbiol Lett 292(2):149–161
5. Redfield RJ (2002) Is quorum sensing a side effect of diffusion sensing? Trends Microbiol 10:365–370
6. Lega J, Passot T (2003). Hydrodynamics of bacterial colonies: a model. Phys Rev E 67(3):031906
7. Ván P, Fülöp T (2005) Weakly non-local fluid mechanics: the Schrödinger equation. Proc R Soc A Math Phys Eng Sci 462(2066):541–557
8. Fuqua WC, Winans SC, Greenberg EP (1994) Quorum sensing in bacteria: the LuxR-LuxI family of cell density-responsive transcriptional regulators. J Bacteriol 176(2):269
9. Hense BA, Kuttler C, Müller J, Rothballer M, Hartmann A, Kreft JU (2007) Does efficiency sensing unify diffusion and quorum sensing?. Nat Rev Microbiol 5(3):230–239
10. West SA, Winzer K, Gardner A, Diggle SP (2012) Quorum sensing and the confusion about diffusion. Trends Microbiol 20(12):586–594
11. Zeldovich YB (1970) Gravitational instability: an approximate theory for large density perturbations. Astron Astrophys 5:84
12. Saluena C, Esipov SE, Poschel T (1997) Powders and grains 97. In: Behringer RP, Jenkins JT (eds) Proceedings of the III international conference on powders and grains. Balkema, Rotterdam, pp 341–344
13. Roy S, Llinas R (2016) Non-local hydrodynamics of swimming bacteria and self-activated process. In: BIOMAT 2015: International symposium on mathematical and computational biology, pp. 153–165
14. Beck M, Wayne CE (2009) Using global invariant manifolds to understand metastability in the Burgers equation with small viscosity. SIAM J Appl Dyn Syst 8(3):1043–1065
15. Long MA, Fee MS (2008) Using temperature to analyse temporal dynamics in the songbird motor pathway. Nature 456(7219):189
16. Llinás RR, Iberall A (1977) A global model of neuronal command-control systems. Biosystems 8(4):233–235

# Electrical Communication Systems in Bacterial Biofilms and Ion-Channels

12

**Abstract**

Besides bacterial quorum sensing mechanism through chemical molecules, bacteria communicate also through electrical signalling mechanism. This electrical communication process is mediated by potassium ion-channels. Here, we explore the reconciliation in bacterial biofilms, metabolic codependence model with extended version, potassium ion-channel based electrical communication systems in bacteria communities, electrical attraction to biofilms and time-sharing behaviour.

## 12.1 Ion-Channels

The presence of ion-channels in bacteria is known for many years, but the functional role of these channels is not clear to us. In 1990s, attention was given to explore $K^+$ channels by crystallographic analysis [1–6]. Researchers investigate the ion-channels from molecular paradigm to uncover the mechanism of ion gating and conductance. Bacterial ion-channels are protein complexes which control ions' flow across cell membrane and have been indispensable medium in long range communication [7]. The structural configuration of bacterial ion-channels has similarity with ion-channels in human brain which gives us a pivotal role of the structural basis of signalling mechanism [2]. Bacteria have several types of ion-channels which include potassium ion-channel KcsA (see in Fig. 12.1) [2], chloride channels [8], sodium channels [9], calcium-gated potassium ion-channels [10] and ionotropic glutamate receptors [11].


S. Majumdar, S. Roy, *Microbial Communication*,
https://doi.org/10.1007/978-981-15-7417-7_12

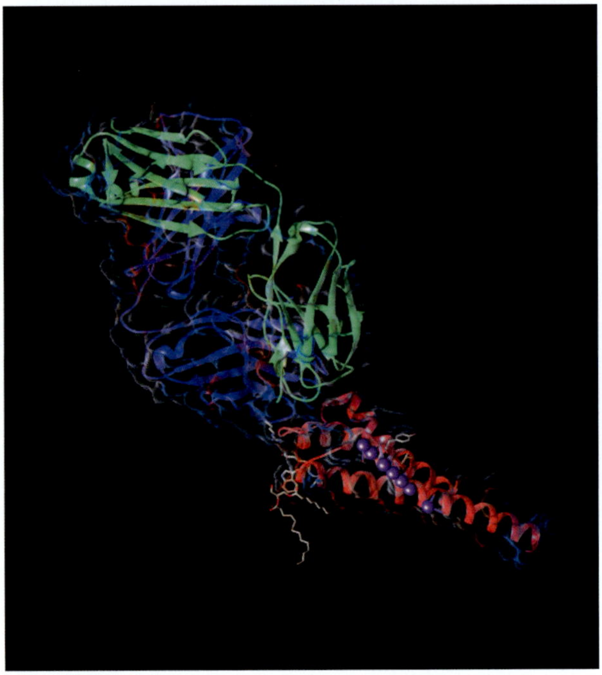

**Fig. 12.1** Illustration of potassium ion-channel from PDB 1K4C

## 12.2 Reconciliation in Bacterial Biofilms

Bacterial cells reside in a community, which we can call as biofilms. Typically, bacterial biofilms develop under nutrient limitation and bacteria within a community compete each other for nutrient. So, the competition among the bacteria is a kind of social behaviour. At the same time bacteria cooperate each other to increase the overall fitness of the bacterial population. The overall fitness depends on the production of common goods and division of labour. When the bacterial communities grow the nutrients supply to interior cells is limited, due the nutrient consumption related with the growth of bacterial cells in biofilm periphery. On the other hand peripheral cells give protection to interior cells from the external attack. So, a fundamental internal conflict is raised between protection and starvation of interior cells in biofilms. The protection of interior cells and opposing benefit of growth are reconciled during biofilm formation [12]. We call it as reconciliation in bacterial biofilms.

This internal conflict at population level is investigated experimentally. Liu et al. [12] sets up an experiment using microfluidics device. They consider *B. subtilis* biofilms for this experimental approach. We summarize experimental observation as follows:

- Oscillation come to light when *B. subtilis* biofilm exceeds a certain colony size.
- Biofilms develop in nutrient limited condition. Metabolism regulates the bacterial growth. Biofilm growth is halted.
- Growth depends on the availability of extracellular ammonium.
- Metabolic codependence between interior and peripheral cells triggers oscillation in biofilm growth that is benefited for the bacterial colony from external attack.
- Metabolic codependence between interior and peripheral cells of biofilm resolve the internal conflict between starvation and protection.

This experimental observations are mathematically formulated in the next sections. Metabolic codependence model with extended version has a good agreement with experimental observations. This study suggests a new strategies to regulate bacterial biofilm growth.

### 12.2.1 Metabolic Codependence Model

Liu et al. [12] proposed the metabolic codependence model, where the nonlinear dynamics of biofilm growth is represented in terms of two distinct bacterial populations, which corresponds to the periphery and interior of the biofilm. Moreover, this two distinct population always keep a constant distance from the physical biofilm edge. Let us consider $G_i$ and $G_p$ as a glutamate concentration in the biofilm interior and periphery, respectively. We assume that the ammonium concentration $A$ is same in both the bacterial populations, because of its fast diffusion. $H_i$ represents the concentration of active glutamate dehydrogenase (GDH) in the interior bacteria cells. Finally, the rate of biomass production is assumed to be given by the concentration of housekeeping proteins in the periphery $r_p$ and interior $r_i$.

The mathematical model is based on the following assumptions:

- The housekeeping proteins production is directly proportional with the concentration of ammonium and glutamate.
- Interior bacterial cells have active GDH.
- GDH activation relies on the availability of glutamate.
- Glutamate and ammonium consumption relies on metabolic activity of bacterial cells.

The dynamic of the metabolic state of the biofilm can be described by the following set of equations

$$\frac{dA}{dt} = \alpha G_i H_i - \delta_A A(r_i + r_p) \tag{12.1}$$

$$\frac{dG_i}{dt} = D(G_p - G_i) - \alpha G_i H_i - \delta_G G_i r_i \tag{12.2}$$

$$\frac{dG_p}{dt} = D(G_i - G_p) + D_E(G_E - G_p) - \delta_G G_p r_p \tag{12.3}$$

$$\frac{dH_i}{dt} = \beta_H \frac{G_i^n}{K_H^n + G_i^n} - \gamma_H H_i \tag{12.4}$$

$$\frac{dr_i}{dt} = \beta_r A G_i - \gamma_r r_i \tag{12.5}$$

$$\frac{dr_p}{dt} = \beta_r A G_p - \gamma_r r_p \tag{12.6}$$

where $\alpha$ is the glutamate dehydrogenation coefficient. $\delta_A$, $\delta_G$, $D$, $D_E$, $G_E$, $\beta_H$, $\gamma_H$, $K_H$, $n$, $\beta_r$ and $\gamma_r$ are the ammonia consumption coefficient, glutamate consumption rate, glutamate diffusion constant with the biofilm, glutamate diffusion constant between biofilm and exterior, glutamate concentration in the external medium, maximal activation rate of GDH, deactivation rate of GDH, GDH activation threshold, Hill coefficient for GDH activation, expression coefficient of ribosomal/housekeeping proteins and degradation rate of ribosomal/housekeeping proteins respectively [12].

Suppose, $\rho$ is the cell density. The dynamics of the cell density of both populations are expressed as

$$\frac{d\rho_{i,p}}{dt} = \eta r_{i,p}\rho_{i,p}\left(1 - \frac{\rho_{i,p}}{K(G_{i,p})}\right) - \lambda_{i,p}\rho_{i,p} \tag{12.7}$$

where $K$ is the carrying capacity which depend on glutamate concentration. $\lambda_i$ and $\lambda_p$ are expansion rate of interior and peripheral cells. We have

$$K(G) = \frac{G^m}{K_k^m + G^m} \tag{12.8}$$

where $K_k$ is the glutamate threshold for carrying capacity. $\eta$ and $m$ are population growth rate coefficient and Hill coefficient for carrying capacity. We say $\mu_{i,p} = \eta r_{i,p}\rho_{i,p}(1 - \frac{\rho_{i,p}}{K(G_{i,p})})$.

### 12.2.2 Extended Model

We add external glutamine to the media and implement into our system as noncompetitive inhibition on two parameters $\alpha$ and $\delta$. We have effective $\bar{\alpha}$ and $\bar{\delta}$ as follows:

$$\bar{\alpha} = \frac{\alpha}{\frac{[Gln]}{K_a} + 1}$$

$$\bar{\delta_{A,G}} = \frac{\delta_{A,G}}{\frac{[Gln]}{K_\delta} + 1}$$

where $[Gln]$ is the concentration of glutamine in the media. $K_a$ and $K_\delta$ are glutamine inhibition threshold on GDH and glutamine synthase activity, respectively.

The addition of ammonium to the media is given by

$$\frac{dA}{dt} = \alpha G_i H_i - \delta_A A(r_i + r_p) + \alpha_0$$

where $\alpha_0$ is the rate of ammonium entering into the biofilm from external medium [12].

Next, we study the effect of overexpressing GDH in biofilm with the little modification in mathematical framework.

$$\frac{dA}{dt} = \alpha G_i H_i + \alpha G_p H_p - \delta_A A(r_i + r_p)$$

$$\frac{dG_p}{dt} = D(G_i - G_p) + D_E(G_E - G_p) - \alpha G_p H_p - \delta_G G_p r_p$$

$$\frac{dH_i}{dt} = \beta_0 + \beta_H \frac{G_i^n}{K_H^n + G_i^n} - \gamma_H H_i$$

$$\frac{dH_p}{dt} = \beta_0 - \gamma_H H_p$$

where $\beta_0$ is the expression rate of GDH from the additional copy of gene.

Finally, we get another modified version of the model when hydrogen peroxide is added to the media [12].

$$\frac{dr_p}{dt} = -\gamma_r r_p$$

$$\frac{d\rho_p}{dt} = \eta r_p \rho_p (1 - \ ) - \lambda_{H_2O_2} \rho_p - \lambda_p \rho_p$$

$$\frac{dH_p}{dt} = -\gamma_H H_p$$

where $\lambda_{H_2O_2}$ is the death rate due to hydrogen peroxide.

Furthermore, we explore the effects of varying the ratio of interior and peripheral cell by the equation

$$\frac{dA}{dt} = f_i \alpha G_i H_i - \delta_A A(f_i r_i + (1 - f_i) r_p)$$

where $f_i$ is the fraction size of interior population over full biofilm population.

## 12.3 Electrical Communication within Bacterial Communities

Biofilm is well organized bacterial community. Billions of densely packet bacteria are lived in biofilms. We already explored bacterial chemical signalling mechanism (or quorum sensing). Now, we investigate the bacterial communication process over long distance. We have noticed an oscillatory behaviour in biofilm which is closely related to long range metabolic codependence between peripheral and interior cells in biofilm. Prindle et al. [13] continued the experiment using microfluidics device and observed the following phenomenon:

- An oscillation in membrane potential is observed.
- Biofilms generate synchronized oscillation in membrane potential.
- Bacterial potassium ion-channels mediate the long range electrical communication within bacterial biofilm communities.
- Signal is propagated as waves of potassium.
- A positive feedback loop generates this wave mechanism. Intercellular potassium is released by metabolic trigger. Finally, depolarization of neighbouring bacterial cells takes place.
- Wave of depolarization regulates metabolic states among bacterial cells in biofilm.

These experimental observations are supported by the mathematical formalism. Now, we describe the mathematical formalism (in the spirit of Hodgkin–Huxley model [14]) to explore our system. The dynamics of the membrane potential $V$ in bacterial cells is given by

$$\frac{dV}{dt} = -g_K n^4 (V - V_K) - g_L (V - V_L) \qquad (12.9)$$

where $g_K$ is the channel conductance.

We assume that potassium channel analogous to Hodgkin–Huxley model and is constructed by four subunits. These are open for a fraction of time $n$ with the dynamics

$$\frac{dn}{dt} = \alpha(S)(1 - n) - \beta n \qquad (12.10)$$

where $\alpha(S) = \frac{\alpha_0 S^m}{S_{th}^m + S^m}$. $\alpha$ is the opening rate of the potassium channel which depends on metabolic stress. $S$ is the concentration of the metabolic stress related metabolic product. When $S$ becomes larger then the channel is opened. Assume, the depolarization of cell membrane controls the stress. Thus, we have

$$\frac{dS}{dt} = \frac{\alpha_S(V_{th} - V)}{\exp\left(\frac{V_{th}-V}{\sigma}\right) - 1} - \gamma_S S \tag{12.11}$$

Let us consider that $E$ be the concentration of excess extracellular potassium and $V_K = V_{K0} + \delta_K E$, $V_L = V_{L0} + \delta_L E$. $E$ increases when potassium channel opens. We have

$$\frac{dE}{dt} = F g_K n^4 (V - V_K) - \gamma_e E \tag{12.12}$$

where $\gamma_e$ is the linear decay rate. $F$ describes the relation between membrane potential and charges in excess extracellular potassium.

Suppose, $T$ is the concentration of ThT. The dynamics of concentration of ThT is expressed as

$$\frac{dT}{dt} = \alpha_t (V_{L0} - V) - \gamma_t T \tag{12.13}$$

where $\gamma_t$ is the decay rate.

We extend the mathematical framework in space to describe the propagation of potassium signal across the biofilm

$$\frac{dE_i}{dt} = F g_K n_i^4 (V_i - V_K) - \gamma_e E_i + \frac{D}{\Delta x^2}(E_{i+1} - E_{i-1} - 2E_i)$$

where $\gamma_e$ is the decay of extracellular potassium. $i$ is the subindex which describes the box along the chain. $D$ is the diffusion coefficient of potassium.

Finally, we model the effect of *yugO* deletion with the leaky potassium outflux form:

$$I_{trkA} = g_{trkA}([K]_{int} - E) = g_{trkA}(f - 1)E$$

where $f$ is the normalized intercellular potassium concentration.

Experimental observation together with mathematical model suggest that $K^+$ channels regulate electrical signal to coordinate metabolism within biofilm [13]. The electrical communication process is established in bacterial communities. It also opens a new avenue of research.

## 12.4 Electrical Attraction to Biofilms

Cell–cell communication mechanism coordinates bacterial behaviour within biofilm communities. Recent experiment shows that potassium ion-channels mediated electrical communication process attract distance cells toward *B. subtilis* biofilm [15]. Experimental observation is also validated with mathematical model. Membrane potential of distant is altered by extracellular potassium which are emitted from the biofilm. As a consequence, it is directing distant cells mortality. Moreover, experiment shows that *P. aeruginosa* is attracted towards *B. subtilis* biofilm. So, a cross-species interaction also takes place. Long range electrical signalling mechanism not only coordinate own behaviour in the biofilm communities but also influence other bacteria [15, 16].

## 12.5 Time-Sharing Behaviour

Liu et al. [17] discovered that two distant *B. subtilis* biofilms are coupled through electrical signalling mechanism and synchronize biofilms growth dynamics under the circumstance of metabolic oscillation. Coupling increases competition by synchronizing demand for limited recourses. Communication and competition govern the synchronization between bacterial biofilms. Nutrient limitation is a very common feature in biological system. It is well known that time-sharing is a strategy where user take turns consuming resources. Biofilms resolve conflicts by switching from in phase to anti-phase. Bacterial biofilms take turns consuming resources. In the experiment, it has been confirmed that distant biofilms coordinate their behaviour to resolve nutrient competition through time-sharing. This is an intelligent strategy to share limited nutrients [17, 18].

## References

1. Schrempf H, Schmidt O, Kümmerlen R, Hinnah S, Müller D, Betzler M, Steinkamp T, Wagner R (1995) A prokaryotic potassium ion channel with two predicted transmembrane segments from Streptomyces lividans. EMBO J 14(21):5170–5178
2. Doyle DA, Cabral JM, Pfuetzner RA, Kuo A, Gulbis JM, Cohen SL, Chait BT, MacKinnon R (1998) The structure of the potassium channel: molecular basis of K+ conduction and selectivity. Science 280(5360):69–77
3. Chang G, Spencer RH, Lee AT, Barclay MT, Rees DC (1998) Structure of the MscL homolog from Mycobacterium tuberculosis: a gated mechanosensitive Ion channel. Science 282(5397):2220–2226
4. Dutzler R, Campbell EB, Cadene M, Chait BT, MacKinnon R (2002) X-ray structure of a ClC chloride channel at 3.0 Åreveals the molecular basis of anion selectivity. Nature 415(6869):287–294
5. Jiang Y, Lee A, Chen J, Cadene M, Chait BT, MacKinnon R (2002) The open pore conformation of potassium channels. Nature 417:523–526
6. Booth IR (2003) Bacterial ion channels. In: Genetic engineering. Springer, Boston, pp 91–111

7. Majumdar S, Pal S (2018) Information transmission in microbial and fungal communication: from classical to quantum. J Cell Commun Signal 12(2):491–502
8. Iyer R, Iverson TM, Accardi A, Miller C (2002) A biological role for prokaryotic ClC chloride channels. Nature 419(6908):715–718
9. Ren D, Navarro B, Xu H, Yue L, Shi Q, Clapham DE (2001) A prokaryotic voltage-gated sodium channel. Science 294(5550):2372–2375
10. Jiang Y, Lee A, Chen J, Cadene M, Chait BT, MacKinnon R (2002) Crystal structure and mechanism of a calcium-gated potassium channel. Nature 417(6888):515–522
11. Chen GQ, Cui C, Mayer ML, Gouaux E (1999) Functional characterization of a potassium-selective prokaryotic glutamate receptor. Nature 402(6763):817–821
12. Liu J, Prindle A, Humphries J, Sagarra MG, Asally M, Dong-yeon LD, Ly S, Ojalvo GJ, Süel MG (2015) Metabolic co-dependence gives rise to collective oscillations within biofilms. Nature 523(7562):550–554
13. Prindle A, Liu J, Asally M, Ly S, Ojalvo GJ, Süel MG (2015) Ion channels enable electrical communication in bacterial communities. Nature 527(7576):59–63
14. Hodgkin AL, Huxley AF (1952) A quantitative description of membrane current and its application to conduction and excitation in nerve. J Physiol 117(4):500–544
15. Humphries J, Xiong L, Liu J, Prindle A, Yuan F, Arjes AH, Tsimring L, Süel MG (2017) Species-independent attraction to biofilms through electrical signaling. Cell 168(1):200–209
16. Majumdar S, Pal S (2017). Cross-species communication in bacterial world. J Cell Commun Signal 11(2):187–190
17. Liu J, Corral MR, Prindle A, Dong-yeon LD, Larkin J, Sagarra GM, Ojalvo GJ, Süel MG (2017) Coupling between distant biofilms and emergence of nutrient time-sharing. Science 356(6338):638–642
18. Majumdar S, Pal S (2017). Bacterial intelligence: imitation games, time-sharing, and long-range quantum coherence. J Cell Commun. Signal 11(3):281–284

# Synthetic Biology and Microbial Communication 13

**Abstract**

Synthetic biology is an interdisciplinary subject where we are talking about the rewiring, re-engineering and design principle of the living systems for understanding the undiscovered underlying mechanism of complex biological systems. Here, we discuss mainly synthetic biological approaches related to bacterial communication systems which include genetic toggle switch, repressilator, synthetic bacterial population control circuit, programme pattern formations, synthetic predator–prey model.

## 13.1 Synthetic Biology

We already discussed the mathematical models and experimental observations of the bacterial communication systems. Now, the question is how much we know about these natural phenomena. How to design this communication system from our current understanding of biological system? If we try to address the above question, we may find some difficulties. Major problem is that the biological phenomena are too complex in nature and we do not have enough knowledge about it. We only know the constituent parts of cell-to-cell communication, but we do not know how these parts are organized and its physiological function[1]. That is way we move onto a next era called synthetic biology. It has been noticed that researchers from different fields have an ongoing interest in synthetic biology, which include biologists, chemist, engineers and re-writers [2].

Synthetic biology is an interdisciplinary subject, which bring together biologist, engineers, physicist and chemist to build biomolecular component, pathways and networks. We use these tools for reprogramming and rewiring organisms [3]. In this way, one builds up biological design principle for better understanding of cell behaviour. Here, circuits are designed from scratch and based on genetic. Our lives will be changed by these re-engineered organisms in future. New discovery leads


S. Majumdar, S. Roy, *Microbial Communication*,
https://doi.org/10.1007/978-981-15-7417-7_13

to cure different bacterial infection diseases, creating cheap drugs and advance our knowledge of microbial communication [3–5]. So, the important challenge behind it is to recognize general, scalable, strategies that accredit fabrication of increasingly complex gene circuits with reliable performance, as well as to construct original and significant technological platforms for the quantitative circuit characterization [2, 5–7].

## 13.2 Genetic Toggle Switch

The gene regulatory circuit can be constructed with any kind of desire properties (i.e. oscillations, multistability) from networks of simple regulatory elements. The genetic toggle switch is designed as a bistable gene circuit (in *E. coli*), based on simple mathematical model predictions. The toggle switch contains two promoters and two repressors. Repressor1/repressor 2 inhibits transcription from promoter1/promoter2 and induced by inducer 1/inducer 2 (see Fig. 13.1). We design the toggle switch in such a way so as to achieve bistable behaviour which is robust in nature (i.e. reveal bistability in a wide range of parameters) as well as the switch does not flip randomly between states. We observe two stable states in the absence of inducers [8]. The possible states are as follows:

- promoter 1 transcribes repressor 2.
- promoter 2 transcribes repressor 1.

Switching is completed successfully by introducing inducer of the presently active repressor. The inducer allows the non-allied repressor to be maximally transcribed prior to it stably represses at fast active promoter [8]. Toggle switches are imposed on *E. coli* plasmids. Now, we move on to toggle model, which defines condition for

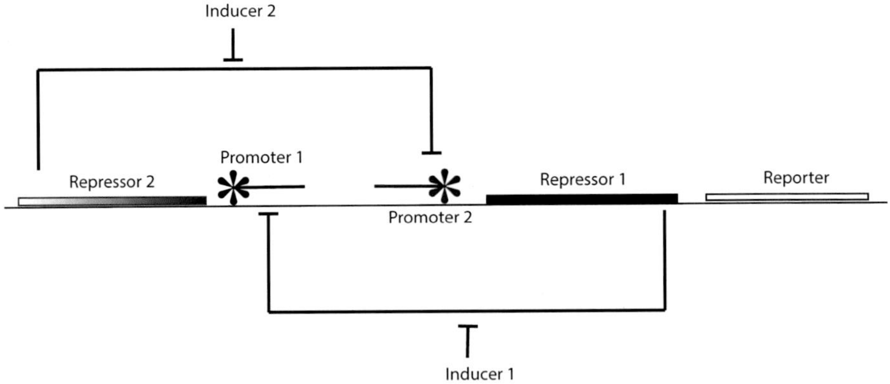


**Fig. 13.1** Schematic diagram of a genetic toggle switch (design)

bistability as well as the toggle switch behaviour. The mathematical expressions are as follows:

$$\frac{du}{dt} = \frac{\alpha_1}{1 + v^{\beta}} - u \tag{13.1}$$

$$\frac{dv}{dt} = \frac{\alpha_2}{1 + u^{\gamma}} - v \tag{13.2}$$

where $u$ and $v$ are the concentrations of repressor 1 and repressor 2, respectively. $\alpha_1$ and $\alpha_2$ represent the effective rate of synthesis of repressor 1 and repressor 2, respectively. $\beta$ and $\gamma$ are the cooperativity of repression of promoter 2 and promoter 1, respectively. The second term of both the equations represents the degradation of repressors. The lump parameters (i.e. $\alpha_1, \alpha_2$) describe the open-complex formation, repressor binding, net effect of RNA polymerase binding, transcript termination, transcript elongation, polypeptide elongation and ribosome binding [8].

We find the nullclines of the toggle model by setting $\dot{u} = 0$ and $\dot{v} = 0$. The origin of the bistability with two stable states and one unstable state is observed as depicted in Fig. 13.2a and b. The key feature of Fig. 13.2a and b is given as follows:

- Bistability of the system is controlled by the cooperative repression of transcription.
- Rates of synthesis of two repressors are balanced.
- Two basins of attraction is created by the toggle network. Hence, toggle starting above the separatrix will settle to state 1 and on the other hand, toggle starting below separatrix ultimately settle to the state 2.

We get the condition for bistability from Fig. 13.2c and d. The size of the bistable region increases when the rates of repressor synthesis are increased. At least one of the inhibitors surely represses expression with cooperativity greater than one. Furthermore, the robustness of the system depends on the higher order cooperativity [8].

The toggle switch is considered as a synthetic addressable cellular memory unit. Toggle makes fundamental basis for genetic applets for control over cell function. It has an application in gene therapy, biocomputing and biotechnology.

## 13.3 Repressilator

Now, we address the design principle of the synthetic network by using three transcriptional repressor systems which are not part of natural biological clock. It is known as repressilator. The repressilator is used to form an oscillating network in *E. coli*. LasI is a first repressor protein that inhibits transcription of *tetR* (second repressor gene) from Tn10 and Tn10 protein product inhibits *cI* (third gene) from $\lambda$ phage. Ultimately, CI inhibits lasI expression. So, we get the complete cycle (see Fig. 13.3). It is a negative feedback loop that generates temporal oscillation

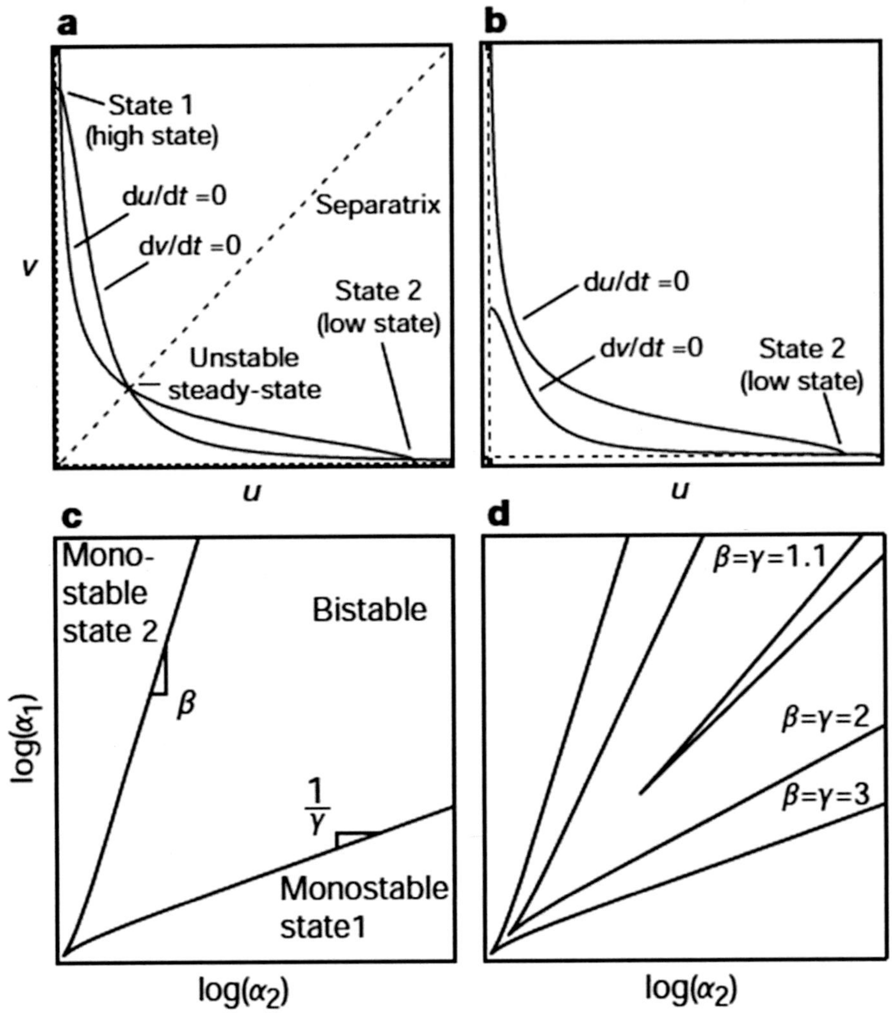


**Fig. 13.2** Toggle model simulation: (**a**) Bistable toggle network, (**b**) Monostable toggle network, (**c**) Bistable region, (**d**) Bifurcation lines are visualized with bistable region (insides each pair of curves) for different values of $\beta$ and $\gamma$ (reproduced with permission from [8])

in concentrations profile of each of its components. We construct a mathematical model of transcriptional regulation and design the repressilator [9].

Let us consider $p_i$ and $m_i$ are the three repressor protein concentrations and their corresponding mRNA concentrations, respectively ($i$ is *lacI*, *tetR* or *cI*).

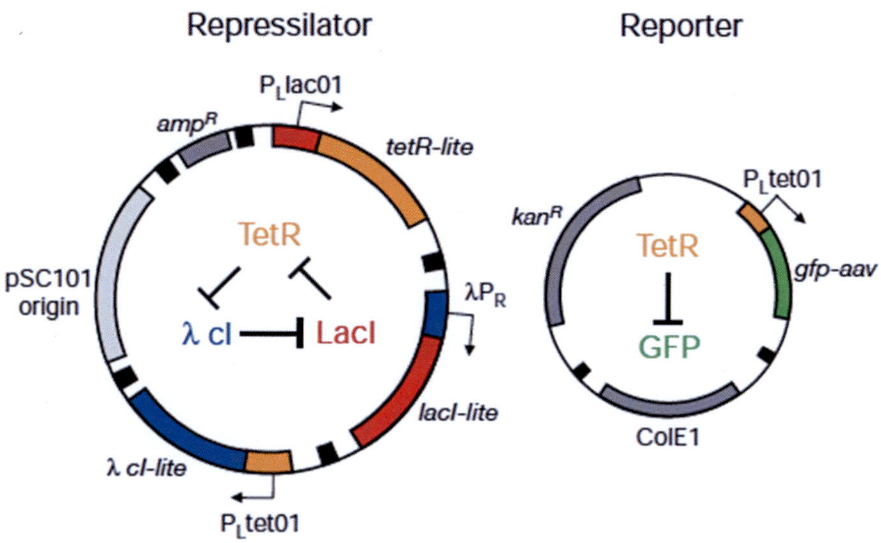


**Fig. 13.3** Repressilator: It is a cyclic negative feedback loop, which is composed by three repressor genes with corresponding promoters (reproduced with permission from [9])

These six molecular species take part in reactions (in translation, transcription and degradation). The kinetics of the system is expressed as

$$\frac{dm_i}{dt} = -m_i + \frac{\alpha}{\left(1 + p_j^n\right)} + \alpha_0 \tag{13.3}$$

$$\frac{dp_i}{dt} = -\beta(p_i - m_i) \tag{13.4}$$

where $i = lacI, tetR, cI$ and $j = cI, lacI, textR$. $\alpha_0$ (in presence of saturated amount of repressor) and $\alpha + \alpha_0$ (in absence of saturated amount of repressor) are the number of protein copies per cell and $n$ is the Hill coefficient. Assume $\beta = \frac{\text{protein decay rate}}{\text{mRNA decay rate}}$ is a ratio [9]. The above system has unique steady state. This steady state becomes unstable when the condition

$$\frac{(\beta + 1)^2}{\beta} < \frac{3X^2}{4 + 2X} \tag{13.5}$$

where $X = -\frac{\alpha n p^{n-1}}{(1+p^n)^2}$ satisfy. Moreover, $p = \frac{\alpha}{1+p^n} + \alpha_0$ is the solution for $p$. We have the limit cycle oscillation in the stability analysis of the steady state. Furthermore, we can go for the stochastic approach and observe the oscillatory behaviour because of stochastic fluctuation of the components [9]. The concept of repressilator is useful to design artificial genetic network and improve our current knowledge of natural oscillatory networks.

## 13.4 Synthetic Bacterial Population Control Circuit

The design of gene circuits inside cells is very difficult because its robust performance deals with noise in gene expression. Here, we focus on the population control circuit, which regulates the density of *E. coli* population. The cell number density is transmitted and detected by elements from bacterial cell-to-cell communication system (quorum sensing) which in turn control the death rate. The proposed circuit controls the bacterial cell number density that is smaller than the limits foisted by the environment. The LuxI protein in LuxI/R system in *V. fischeri* synthesizes acyl-homoserine lactone (AHL). The AHL (autoinducer) accumulates in inside cells and experimental medium as the bacterial cell number density increases. When the AHL concentration reaches the threshold level, it binds and triggers the LuxR transcriptional regulator. Finally, killer gene ($E$) is induced by the active LuxR ( $R^*$) under the influence of $p_{luxI}$ (*luxI* promoter) [10]. As a consequence cell death occurs (see Fig. 13.4). The cell death happens because of sufficient amount of killer protein.

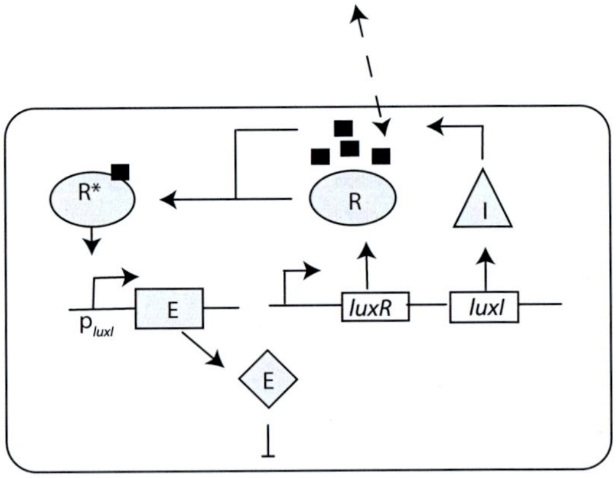


**Fig. 13.4** Schematic visualization of the circuit. $E$ represents killer gene. $R$, $I$ and $R^*$ are LuxR, LuxI and LuxR active, respectively. Filled box is AHL

Now, we express mathematically all major kinetic events in the proposed circuit design, which includes cell growth, death, killer protein production and degradation, AHL signal [10]. We accumulate all the assumptions as follows:

- AHL and degradation of killer protein follows 1st order kinetics with $d_E$( rate constants).
- Cell density $N$ is changing according to logistic kinetics with growth rate $k$ and carrying capacity $N_m$ ( without circuit).
- Intercellular concentration of killer protein $E$ is proportional to $N$ with rate constant $d$(for circuit regulated growth).
- AHL concentration $(A)$ is proportional to production rate of $E$ with $k_E$ (rate constant).
- The production rate of AHL is proportional to $N$ with $\nu_A$ (rate constant).

Hence, we have

$$\frac{dN}{dt} = kN(1 - N/N_m) - dEN \tag{13.6}$$

$$\frac{dE}{dt} = k_E A - d_E E \tag{13.7}$$

$$\frac{dA}{dt} = \nu_A N - d_A A \tag{13.8}$$

Many intermediated steps are lumped in the production term $(k_E A)$ of the Eq. 13.7, which includes binding of activated LuxR to $p_{luxI}$, expression of killer gene and activation of LuxR by AHL. We get that the AHL diffusion is faster than other process. The Eq. 13.6 becomes $\frac{dN}{dt} = N(k - dE)$ when $N \ll N_m$. Then, we have a simplified mathematical model with two steady state solutions. Linear stability analysis gives us

- Trivial steady state $(N_s = 0, E_s = 0, A_s = 0)$ is unstable $\forall$ positive parameters (subscript "s" denotes steady state).
- Nontrivial steady state $(N_s = \frac{d_A d_E k}{\nu_A k_E d}, E_s = k/d, A_s = \frac{d_E k}{k_E d})$ is stable iff $d_A + d_E > k$.

Finally, it can be shown that the nontrivial steady state is stable $\forall$ biologically feasible parameters. The mathematical model predicts that proposed circuit sets a stable steady state in terms of gene expression and cell density, which is tunable by altering cell communication signal. Experiments confirm the mathematical prediction of damped oscillation while approaching steady state. We consider this synthetic approach as a foundation of bacterial communication programme inside the communities which includes quorum sensing regulated growth and death concept [10]. It is the beginning of a new paradigm called engineering synthetic ecosystems.

## 13.5 Programme Pattern Formations

We usually consider pattern formation as a hallmark feature of coordinated cell behaviour in case of multicellular and signal organisms. So, pattern formation is also expected to occur in intra- and intercellular bacterial communication systems. It is possible to design receiver cells and synthetic multicellular systems to create ring like patterns of differentiation. How receivers at intermediate distance from sender generate the output protein is systematically visualized in Fig. 13.5.

Expression of LuxI enzyme initiates the communication process from senders. AHL synthesis is catalyzed by LuxI. A chemical gradient is created around the senders by the diffused AHL. It has been observed that AHL diffuses into close vicinity by receiver cells. Next it follows several processes which include LuxR bound, transcriptional regulator (AHL dependent), activation of expression of lambda (CI) and Lac repressor. High concentration of AHL is received when the receiver and sender cells are in close vicinity. We observe high cytoplasmic levels of $LacI_{M1}$ and CI and GFP repression as a consequence. Low concentration of AHL is achieved when the receiver and sender cells are far away. As a result, CI, $LacI_{M1}$ (basel level) and wild type LacI are expressed and GFP repression is observed. Finally, intermediate AHL concentration is reached when the receiver and sender cells are at intermediate distances. CI shuts off LacI expression because the repression efficiency of $LacI_{M1}$ is significantly lower than CI. Furthermore, a condition, i.e. the concentration of $LacI_{M1}$ is below the threshold level is required for GFP repression [11].

Now, we discuss the mathematical model to express single bacterial cell band detected response to AHL. It is also possible to monitor the spatiotemporal pattern of sender–receiver multicellular system using dynamical system. Let us consider LacI ($L$), AHL ($A$), LuxR/AHL complex ($R$), CI ($C$), GFP ($G$) as five intercellular species [11]. The concentration of LuxR is fixed. The following mathematical expression is used

$$\frac{dG}{dt} = \frac{\alpha_G}{1 + (L/\beta_L)^{\eta 1}} - \gamma_G G \tag{13.9}$$

$$\frac{dL}{dt} = \frac{\alpha_{L1}}{1 + (C/\beta_C)^{\eta 2}} + \frac{\alpha_{L2} \cdot R^{\eta 3}}{(\theta_R)^{\eta 3} + R^{\eta 3}} - \gamma_L L \tag{13.10}$$

$$\frac{dC}{dt} = \frac{\alpha_C R^{\eta 3}}{(\theta_R)^{\eta 3} + R^{\eta 3}} - \gamma_C C \tag{13.11}$$

$$\frac{dR}{dt} = \rho_R [LuxR]^2 A^2 - \gamma_R R \tag{13.12}$$

$$\begin{aligned}\frac{dA_{x,y,z}}{dt} &= \zeta\big(A_{x-1,y,z} + A_{x+1,y,z} + A_{x,y-1,z} + A_{x,y+1,z} \\ &\quad + A_{x,y,z-1} + A_{x,y,z+1} - 6A_{x,y,z}\big) - \gamma_A\end{aligned} \tag{13.13}$$

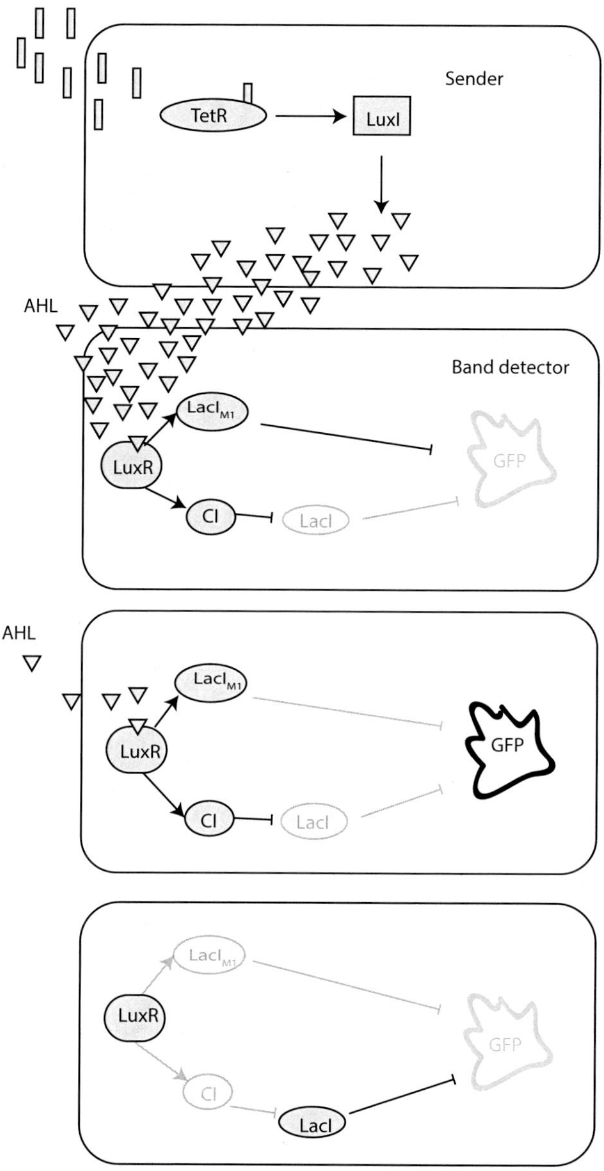


**Fig. 13.5** Schematic visualization of bacterial multicellular system

**Table 13.1** List of parameters used in the model

| Variables | Descriptions |
|---|---|
| $\alpha_G, \alpha_{L1}, \alpha_{L2}, \alpha_C$ | Protein synthesis rates |
| $\beta_L, \beta_C$ | Repression coefficients |
| $\gamma_C, \gamma_L, \gamma_G, \gamma_R$ | Protein decay |
| $\theta_R$ | LuxR/AHL activation coefficient |
| $\eta 1, \eta 2, \eta 3$ | Transcription factor cooperativity |
| $\rho_R$ | LUX/AHL dimerization |
| $\zeta$ | AHL intercellular diffusion |
| $\gamma_A$ | AHL decay |

The above-mentioned mathematical formalism together with experiment shows distinct ring like patterns. Hearts, ellipses and clovers patterns are also achieved by placing the sender cells in various configurations. So, this design principle helps us to construct artificial multicellular system [11]. This system shows programmed pattern formation indeed which is useful for biosensing, tissue engineering and biomaterial fabrication as well (Table 13.1).

## 13.6 A Synthetic Predator–Prey Model

The synthetic predator–prey model is the construction of a synthetic ecosystem. It is a very challenging task because predator usually produces rich dynamics in the system. The synthetic predator–prey is constructed by a gene circuit (synthetic) which converts two *E. coli* populations into system. Predator and prey communicate via bacterial density dependent collective behaviour called quorum sensing. This bi-directional communication controls each other gene expression. LuxI/R and LasI/r systems are used for communication. Predators cells are dying at the low prey cell density, due to expression of *ccdB* (suicide gene). In prey cells, AHL (3OC6HSL) is synthesized by LuxI. When the prey cells density increases, AHL accumulation takes place and the AHL concentration becomes sufficiently high to trigger expression of *ccda* (antidote gene) for rescue predator cells. Predator cells generate another AHL (3OC12HSL) by LasI in succession. As a consequence it activates the expression of *ccdB* (killer gene), inducing predation. The system differs from the canonical predator–prey system because first, the prey rescues predator by antidote protein and second, competition for nutrient is appeared in the system [12].

Now, we mathematically formulate the synthetic *E. coli* predator–prey system based on the following major assumptions:

- No crosstalk is allowed between different AHL signals.
- Mass action kinetics is followed in case of (1) AHL is binding with cognate regulator, (2) active regulator dissociation and (3) active regulator dimerization processes.
- AHL is transported through cell membrane.

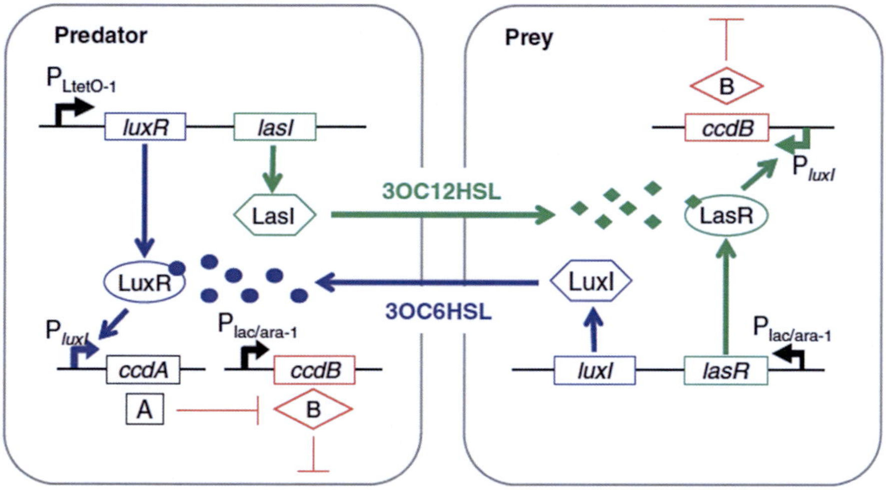


**Fig. 13.6** Schematic visualization of synthetic predator–prey system (adapted from [12])

- Each and every AHL has uniform concentration in well mixed medium and inside the cell.
- The concentration difference between intra- and extracellular space is proportional to the flux of AHL across cell membrane.
- Michaelis–Menten kinetics is followed by the regulation of *ccdB* killer gene expression.
- AHL synthesis occurs at constant rate.
- Bacterial cell growth follows logistic kinetics. Assume $\hat{k}_{ci}$ (for $i = 1$ (predator) and $i = 2$ (prey)) be specific growth rate and $\hat{c}_{\max}$ be carrying capacity for mixture (predator and prey). $\hat{d}_i$ is rate constant in case of cell death (Fig. 13.6).

Assume $\hat{c}_1$ and $\hat{c}_2$ be predator and prey cells, respectively. We express cell growth and death as follows:

$$\frac{d\hat{c}_1}{dt} = \hat{k}_{c1}\hat{c}_1\left(1 - \frac{\hat{c}_1 + \hat{c}_2}{\hat{c}_{\max}}\right) - \frac{\hat{d}_1\hat{E}_1\hat{c}_1}{1 + \hat{\alpha}_A\hat{A}} - D\hat{c}_1 \tag{13.14}$$

$$\frac{d\hat{c}_2}{dt} = \hat{k}_{c2}\hat{c}_2\left(1 - \frac{\hat{c}_1 + \hat{c}_2}{\hat{c}_{\max}}\right) - \hat{d}_2\hat{E}_2\hat{c}_2 - D\hat{c}_2 \tag{13.15}$$

Mathematical expressions of regulator genes $\hat{M}_{Ri}$ and decay of products $\hat{R}_i$ are

$$\frac{d\hat{M}_{Ri}}{dt} = \hat{\nu}_{MRi} - \hat{d}_{MRi}\hat{M}_{Ri} \tag{13.16}$$

$$\frac{d\hat{R}_i}{dt} = \hat{k}_{Ri}\hat{M}_{Ri} - \hat{d}_{Ri}\hat{R}_i - \hat{k}_{Pi}\hat{A}_{aj}\hat{R}_i + \hat{d}_{Pi}\hat{P}_i \tag{13.17}$$

Assume $\hat{P}_1 = \hat{R}_1 - \hat{A}_{a2}$ and $\hat{P}_2 = \hat{R}_2 - \hat{A}_{a2}$. Now activation of regulator inducer complex is expressed as

$$\frac{d\hat{P}_i}{dt} = \hat{k}_{Pi}\hat{A}_{aj}\hat{R}_i - \hat{d}_{Pi}\hat{P}_i \tag{13.18}$$

Next, we formulate the expression of antidote gene $\hat{M}_A$ and protein in predator $\hat{A}$ as

$$\frac{d\hat{M}_A}{dt} = \frac{\hat{k}_{MA}\hat{\alpha}_{MA}\hat{P}_1^{\beta}}{1+\hat{\alpha}_{MA}\hat{P}_1^{\beta}} - \hat{d}_{MA}\hat{M}_A \tag{13.19}$$

$$\frac{d\hat{A}}{dt} = \hat{k}_A\hat{M}_A - \hat{d}_A\hat{A} \tag{13.20}$$

Then move onto the formation of decay of products $\hat{E}_1$, $\hat{E}_2$ and expression of killer genes $\hat{M}_{E1}$, $\hat{M}_{E2}$

$$\frac{d\hat{M}_{E1}}{dt} = \hat{\alpha}_{ME1} - \hat{d}_{ME1}\hat{M}_{E1} \tag{13.21}$$

$$\frac{d\hat{M}_{E2}}{dt} = \frac{\hat{k}_{ME2}\hat{\alpha}_{ME2}\hat{P}_2^{\beta}}{1+\hat{\alpha}_{ME2}\hat{P}_2^{\beta}} - \hat{d}_{ME2}\hat{M}_{E2} \tag{13.22}$$

$$\frac{d\hat{E}_i}{dt} = \hat{k}_{Ei}\hat{M}_{Ei} - \hat{d}_{Ei}\hat{E}_i \tag{13.23}$$

Finally, we have the product, decay and diffusion of AHLs

$$\frac{d\hat{A}_i}{dt} = \hat{\nu}_{Ai} - \hat{\eta}_i(\hat{A}_i - \hat{A}_{ei}) - \hat{d}_{Ai}\hat{A}_i \tag{13.24}$$

$$\frac{d\hat{A}_{ei}}{dt} = \hat{\eta}_i\frac{\hat{c}_i}{1-\hat{c}_1-\hat{c}_2}(\hat{A}_i - \hat{A}_{ei}) - \hat{\eta}_i\frac{\hat{c}_j}{1-\hat{c}_1-\hat{c}_2}(\hat{A}_{ei} - \hat{A}_{ai}) - \hat{d}_{Aei}\hat{A}_{ei} - D\hat{A}_{ei} \tag{13.25}$$

$$\frac{d\hat{A}_{ai}}{dt} = \hat{\eta}_i(\hat{A}_{ei} - \hat{A}_{ai}) - \hat{d}_{Aai}\hat{A}_{ai} - \hat{k}_{Pj}\hat{A}_{ai}\hat{R}_j + \hat{d}_{Pj}\hat{P}_j \tag{13.26}$$

**Table 13.2** List of state variables used in the model

| Variables | Descriptions |
|---|---|
| $\hat{c}_i$ | Cell density |
| $\hat{E}_i$ | Concentration of *CcdB* Killer protein |
| $\hat{M}_{Ei}$ | Concentration of killer mRNA |
| $\hat{A}_i$ | Concentration of AHL in source cell |
| $\hat{A}_{ei}$ | Concentration of AHL in the medium |
| $\hat{A}_{ai}$ | Concentration of AHL in target cell |
| $\hat{R}_i$ | Concentration of regulator |
| $\hat{M}_{Ri}$ | Concentration of regulator mRNA |
| $\hat{P}_i$ | Concentration of AHL- regulator complex |
| $\hat{Q}_i$ | Concentration of AHL-regulator Complex-2 |

This completes the mathematical formalism of the synthetic predator–prey system. We accumulate the parameters with descriptions in the Table 13.2.

This mathematical model can be simplified by assuming that some components (i.e. all mRNAs, killer proteins, transcriptional regulators) are in quasi steady state. The quantitative nature of the dynamical system is not significantly effective by the assumptions [12]. Hence we have

$$\frac{d\hat{c}_1}{dt} = \hat{k}_{c1}\hat{c}_1\Big(1 - \frac{\hat{c}_1 + \hat{c}_2}{\hat{c}_{\max}}\Big) - \hat{d}_{c1}\hat{c}_1\frac{\hat{K}_1}{\hat{K}_1 + \hat{A}_{e2}^{\hat{\beta}}} - D\hat{c}_1 \tag{13.27}$$

$$\frac{d\hat{c}_2}{dt} = \hat{k}_{c2}\hat{c}_2\Big(1 - \frac{\hat{c}_1 + \hat{c}_2}{\hat{c}_{\max}}\Big) - \hat{d}_{c2}\hat{c}_2\frac{\hat{A}_{e1}^{\hat{\beta}}}{\hat{K}_2 + \hat{A}_{e1}^{\hat{\beta}}} - D\hat{c}_2 \tag{13.28}$$

$$\frac{d\hat{A}_{e1}}{dt} = \hat{k}_{A1}\hat{c}_1 - \big(\hat{d}_{Ae1} + D\big)\hat{A}_{e1} \tag{13.29}$$

$$\frac{d\hat{A}_{e2}}{dt} = \hat{k}_{A2}\hat{c}_2 - \big(\hat{d}_{Ae2} + D\big)\hat{A}_{e2} \tag{13.30}$$

where $\hat{K}_1 = \frac{1}{(\hat{\alpha}_A \frac{\hat{k}_A}{\hat{d}_A}\frac{\hat{k}_{MA}}{\hat{d}_{MA}} + 1)\hat{\alpha}_{MA}(\frac{\hat{k}_{p1}}{\hat{d}_{p1}}\frac{\hat{k}_{R1}}{\hat{d}_{R1}}\frac{\hat{\upsilon}_{MR1}}{\hat{d}_{MR1}})^{\hat{\beta}}}$, $\hat{K}_2 = \frac{1}{\hat{\alpha}_{ME2}\hat{a}_2^{\hat{\beta}}}$, $\hat{d}_{c1} = \hat{d}_1\frac{\hat{k}_{E1}}{\hat{d}_{E1}}\frac{\hat{\alpha}_{ME1}}{\hat{d}_{ME1}}$ and $\hat{d}_{c2} = \hat{d}_2\frac{\hat{k}_{E2}}{\hat{d}_{E2}}\frac{\hat{k}_{ME2}}{\hat{d}_{ME2}}$ (Table 13.3).

Moreover, we have the stochastic version of the above mathematical expressions as follows:

$$\frac{d\hat{c}_1}{dt} = \hat{k}_{c1}\hat{c}_1\Big(1 - \frac{\hat{c}_1 + \hat{c}_2}{\hat{c}_{\max}}\Big) - \hat{d}_{c1}\hat{c}_1\frac{\hat{K}_1}{\hat{K}_1 + \hat{A}_{e2}^{\hat{\beta}}} - D\hat{c}_1 + \hat{\epsilon}\cdot\hat{\zeta} \tag{13.31}$$

**Table 13.3** List of parameters used in the model

| Variables | Descriptions |
|---|---|
| $\hat{k}_i$ | Specific cell growth rate constant |
| $\hat{c}_{\max i}$ | Carrying capacity for cell growth |
| $\hat{d}_i$ | Cell death rate constant |
| $\hat{k}_{Ei}$ | *CcdB* killer protein synthesis rate constant |
| $\hat{d}_{Ei}$ | *CcdB* killer protein decay rate constant |
| $\hat{k}_{MEi}$ | Maximal rate of *ccdB* killer gene transcription |
| $\hat{\alpha}_{MEi}$ | Sensitivity of *ccdB* killer gene transcription to AHL |
| $\hat{d}_{MEi}$ | Killer mRNA decay rate constant |
| $\hat{\upsilon}_{MRi}$ | Transcription rate for a regulator |
| $\hat{d}_{MRi}$ | Regulator decay rate constant |
| $\hat{\upsilon}_{Ai}$ | AHL synthesis rate constant |
| $\hat{\eta}_i$ | AHL diffusion rate constant across the cell membrane |
| $\hat{d}_{Ai}$ | AHL intracellular decay rate constant |
| $\hat{d}_{Aei}$ | AHL extracellular decay rate constant |
| $\hat{k}_{Pi}$ | AHL/regulator binding rate constant |
| $\hat{d}_{Pi}$ | AHL/regulator dissociation rate constant |
| $\hat{k}_{Qi}$ | AHL-regulator complex dimerization rate constant |
| $\hat{d}_{Qi}$ | AHL-regulator-2 unbinding rate constant |
| $\hat{k}_{c1}$ | Predator cell growth rate constant |
| $\hat{k}_{c2}$ | Prey cell growth rate constant |
| $\hat{c}_{\max}$ | Carrying capacity for cell growth |
| $\hat{\beta}$ | Cooperativity of AHL effect |
| $\hat{d}_{c2}$ | Prey cell death rate constant |
| $\hat{K}_i$ | Concentration of AHL necessary to half maximally active PluxI promoter |
| $\hat{k}_{Ai}$ | Synthesis rate constant of AHL by the predator cell |
| $\hat{d}_{Ae1}$ | Decay rate constant of 3OC12HsL in the cell |
| $\hat{d}_{Ae2}$ | Decay rate constant of 3OC6HSL in the cell |
| $D$ | Dilution rate |

$$\frac{d\hat{c}_2}{dt} = \hat{k}_{c2}\hat{c}_2\left(1 - \frac{\hat{c}_1 + \hat{c}_2}{\hat{c}_{\max}}\right) - \hat{d}_{c2}\hat{c}_2\frac{\hat{A}_{e1}^{\hat{\beta}}}{\hat{K}_2 + \hat{A}_{e1}^{\hat{\beta}}} - D\hat{c}_2 + \hat{\epsilon}\cdot\hat{\zeta} \tag{13.32}$$

$$\frac{d\hat{A}_{e1}}{dt} = \hat{k}_{A1}\hat{c}_1 - \left(\hat{d}_{Ae1} + D\right)\hat{A}_{e1} + \hat{\epsilon}\cdot\hat{\zeta} \tag{13.33}$$

$$\frac{d\hat{A}_{e2}}{dt} = \hat{k}_{A2}\hat{c}_2 - \left(\hat{d}_{Ae2} + D\right)\hat{A}_{e2} + \hat{\epsilon}\cdot\hat{\zeta} \tag{13.34}$$

where $\hat{\epsilon}$ is the noise amplitude and $\hat{\zeta}$ is the noise term which satisfies normal distribution with unit variance and zero mean.

Synthetic predator–prey model gives us a framework of synthetic ecosystem model in terms of complexity of the system and provides a program for assembling cellular component to coordinate complex behaviour. Mathematical model captures the oscillatory dynamics of the system which is also explored by experimental observations. We explore the impact of the dilution rate on the synthetic system. Oscillatory dynamics and coexistence of predator–prey is depended on operating conditions in microchemostats. It also allows us to manipulate different intrinsic parameters such as death rate, growth rate, strength of communication [12]. The study allows us to investigate the interplay between population dynamics, gene regulation and environment.

## 13.7 Model of Quorum Sensing Gene Clock

Synthetic biology is an expanding filed which has interdisciplinary nature. We discussed genetic toggle switch, repressilator, programme pattern formations, synthetic predator–prey model and synthetic bacterial population control circuit. Now, we introduce the concept of synchronized clocks which has significance in the coordination of rhythmic behaviour among single elements in a large complex system. The elements of the quorum sensing mechanisms in *V. fisheri* and *B. thuringiensis* are used to design synchronized oscillator (see Fig. 13.7). The production of three genes (*luxI*, *aiiA* and *yemGFP*) is driven by *luxI* promoter in three identical transcriptional modules. AHL is generated from LuxI synthase and diffuses through cell membrane. AiiA negatively regulates the promoter. This gene network has intercellular coupling which generates synchronized oscillations in a growing bacterial population [13].

We investigate the synchronization properties as well as the spatiotemporal wave dynamics using microfluidic devices. We generate the experimentally observed properties using mathematical model. Let us consider that $H_e$, $H_i$, $A$ and $I$ are the concentration of external AHL, internal AHL, AiiA and LuxI. The mathematical

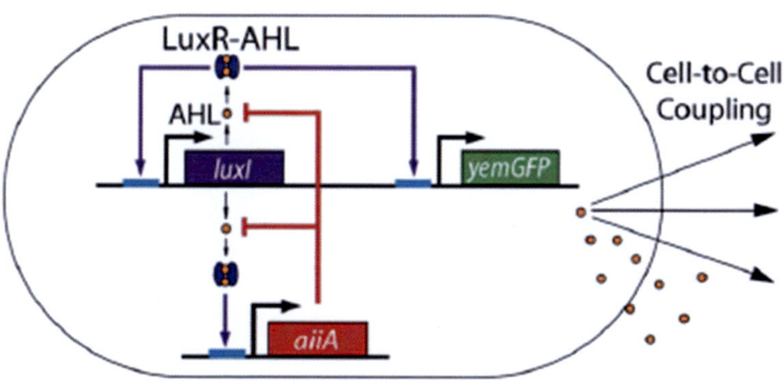


**Fig. 13.7** Network diagram of synchronized genetic clock (reproduced with permission from [13])

model of quorum sensing gene clock is given by

$$\frac{\partial A}{\partial t} = C_A\left[1 - (d/d_0)^4\right]P(\alpha, \tau) - \frac{\gamma_A A}{1 + f(A + I)} \tag{13.35}$$

$$\frac{\partial I}{\partial t} = C_I\left[1 - (d/d_0)^4\right]P(\alpha, \tau) - \frac{\gamma_I I}{1 + f(A + I)} \tag{13.36}$$

$$\frac{\partial H_i}{\partial t} = \frac{bI}{1 + kI} - \frac{\gamma_H A H_i}{1 + gA} + D(H_e - H_i) \tag{13.37}$$

$$\frac{\partial H_e}{\partial t} = -\frac{d}{1 - d}D(H_e - H_i) - \mu H_e + D_1\frac{\partial^2 H_e}{\partial x^2} \tag{13.38}$$

where $P(\alpha, \tau) = \frac{\delta + \alpha H_\tau^2}{1 + k_1 H_\tau^2}$ is the Hill function which describes the delayed generation of corresponding proteins. $H_\tau(t) = H_i(t - \tau)$ represents that $P(\alpha, \tau)$ depends on the past internal AHL concentration. The slowing down process of protein synthesis is given by the pre factor $[1 - (d/d_0)^4]$. The enzymatic degradation of AHL and proteins is described by the terms proportional to $\gamma_x$. The diffusion of AHL is represented by the terms proportional to $D$. $D_1$ is the diffusion constant. Dilution of external AHL is described by the term proportional to $\mu$. The cell density is $d$. This model follows the Michaelis–Menten kinetics [13].

This mathematical model explains the spatiotemporal quorum clock dynamics in nontrivial sense. Here, extracellular AHL is used for coupling among genetic clocks in different cells. The numerical simulation shows synchronized oscillations in cell population (see Figs. 13.8 and 13.9). Coherence of oscillations is increasing with $D_1$. Middle cell begins to oscillate when $D_1 = 0$ (absence of AHL diffusion) and all other bacterial cells are inactive. When $D_1 \neq 0$ (diffusion is present), bacterial cells influence other bacterial cells in the vicinity and propagate oscillation in travelling waveform. The quantitative nature of the oscillation has a good agreement with experimental observations [13].

The bacterial quorum sensing process is used to couple genetic clocks. This finally leads to synchronized oscillation in the colony. This design of networks can be modified in future to explain the complex dynamics of multicellular population and biosensor with oscillatory output [13].

## 13.8 Communication Between Natural and Artificial Cells

Artificial bacterial cells are engineered cells which mimic cellular life. The making of artificial cells gives attention on self-replicating system which is a significant feature of life. But it is insufficient to evaluate how lifelike a biochemical system is. In a recent advancement in the field, researchers reconstruct a well characterized quorum sensing pathways to make artificial cells which have a capability to mimic

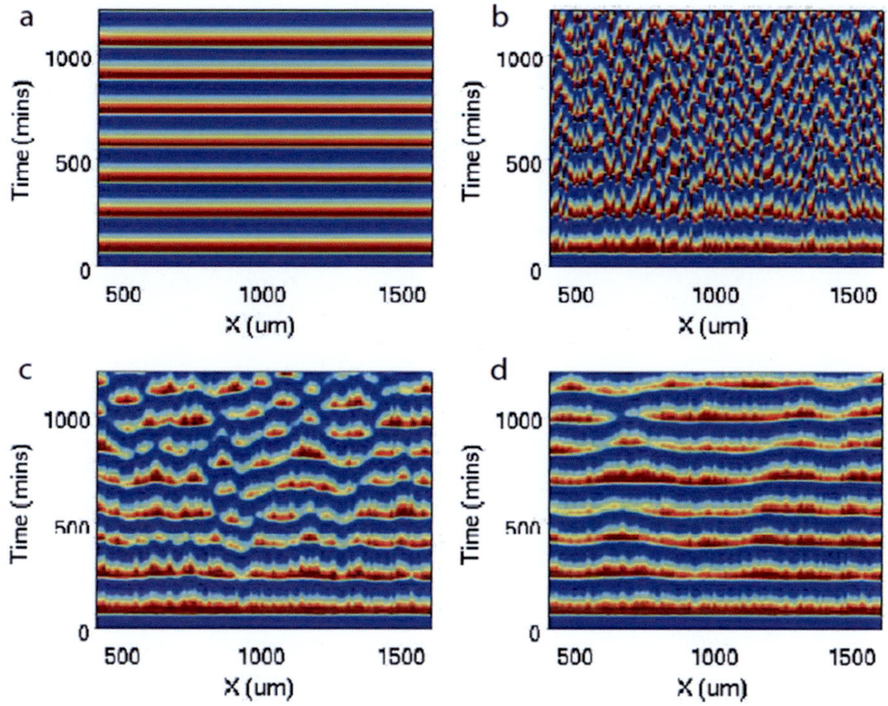


**Fig. 13.8** Numerical simulation of mathematical model shows synchronization of oscillation in the spatially extended system with diffusion. The parameter $p = p_0(1 + \eta\zeta)$ is varying, where $\zeta$ is random number and $\zeta \in [-0.5, 0.5]$. $\eta$ denotes fluctuations magnitude. We vary $D_1$ and $\eta$ in the simulation such as (**a**) ($\eta = 0$ and $D_1 = 0$), (**b**) ($\eta = 0.1$ and $D_1 = 0$), (**c**) ($\eta = 0.1$ and $D_1 = 800\,\mu m^2/s$), (**d**) ($\eta = 0.1$ and $D_1 = 4000\,\mu m^2/s$) (reproduced with permission from [13])

natural cells. Artificial cells are sending and sensing chemical signalling molecules. They are able to talk with natural cells of *V. fisheri*, *E. coli* and *P. aeruginosa*. This activity is judged by RNA-seq, RT-qPCR, fluorescence and luminescence [14].

These experimental observations have a possible application in cellular Turing test. Turing mentioned that the capability of the machine to deceive a judge through textual communication into believing that the machine is a person was used to circumvent the problem of defining intelligence [15]. So, the ability of an artificial cell to deceive a natural cell can evaluate the proper artificial cell. We think about the possible cellular Turing test because all cells communicate from chemical communication pathways in bacteria to pheromone responses in higher organism [14].

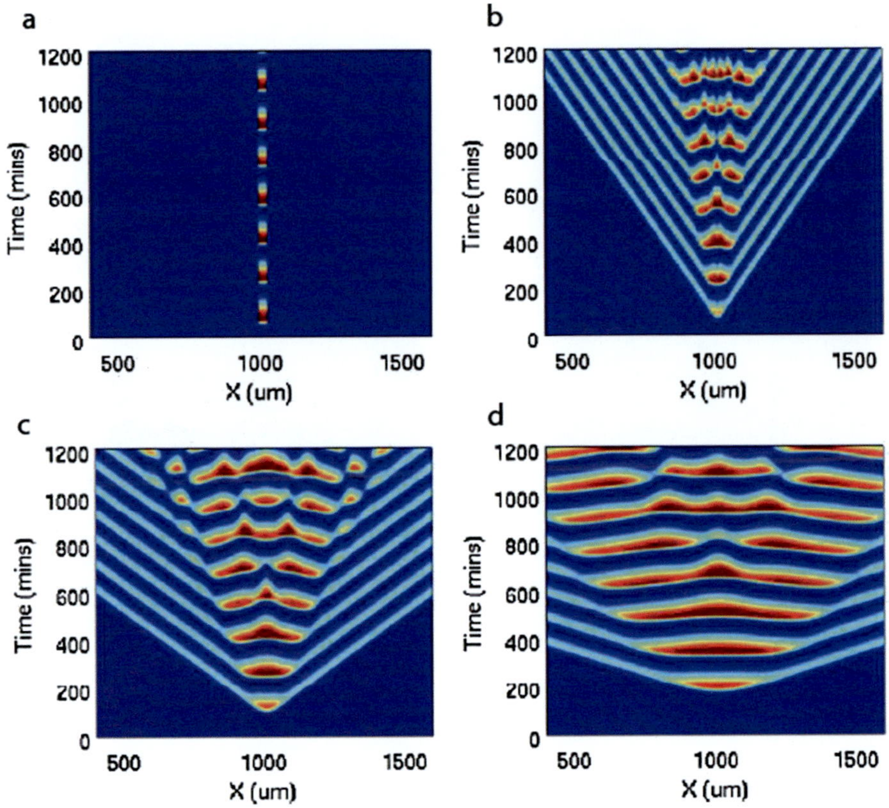


**Fig. 13.9** Wave propagation is presented in spatial uniform system with several external AHL diffusion rates. (**a**) $D_1 = 0$, (**b**) $D_1 = 200\,\mu m^2/s$, (**c**) $D_1 = 800\,\mu m^2/s$ and (**d**) $D_1 = 4000\,\mu m^2/s$ (reproduced with permission from [13])

## 13.9 Stochastic Turing Patterns in Bacterial Population

A fundamental question in science is how patterns are originated and emerged from initial state. We are trying to search the answer of these complex problems from Alan Turing point of view of pattern formation. Alan Turing showed in his seminal paper "The chemical basis of morphogenesis" that chemical morphogenesis could arise from linear instability of spatially uniform state. The periodic pattern formations are shown in reaction diffusion systems with a condition (i.e. activator morphogen diffuses slower than inhibitor morphogen) [16]. Actually, Turing condition is very hard to achieve in natural living systems. Later, the stochastic activator–inhibitor system is developed that predicts patterns formation over a wide range of parameters without strong condition. We explore this theoretical prediction in synthetic bacterial population. The synthetic bacterial population has collective interaction and well characterized bacterial communication mechanism.

The synthetic gene network is designed by two artificial diffusible morphogens (i.e. A3OC12HSL and IC4HSL) from *P. aeruginosa* quorum sensing systems (i.e. *las* and *rhl* system). A3OC12HSL is activator and IC4HSL is inhibitor in the system. This synthetic system exhibits disordered patterns with tunable features on spatial scale [17]. This biologically complex phenomenon is guided by the mathematical framework.

Let $U$ and $V$ be the concentration of A3OC12HSL and IC4HSL, respectively. Assume, $I_u$ and $I_v$ be the concentration of corresponding AHL synthesis, respectively. $C$ denotes to CI. We assume that protein half-life is much longer than mRnA half-life. The protein degradation rate is much slower than operator states of promoter fluctuation. We assume that $\alpha_u$, $\alpha_v$, $\alpha_{iu}$, $\alpha_{iv}$, $\alpha_c$ are A3OC12HSL production rate, IC4HSL production rate, basal production rate of LasI, basal production rate of Rhll and basal production rate of CI, respectively. $\gamma_u$, $\gamma_v$, $\gamma_{iu}$, $\gamma_{iv}$, $\gamma_c$ represent A3OC12HSL, IC4HSL, LasI, Rhll and CI degradation rate, respectively. $D_u$ and $D_v$ are A3OC12HSL, IC4HSL diffusion coefficient [17]. The deterministic reaction–diffusion model is given by

$$\frac{\partial U}{\partial t} = \alpha_u I_u - \gamma_u U + D_u \nabla^2 U \tag{13.39}$$

$$\frac{\partial V}{\partial t} = \alpha_v I_v - \gamma_v V + D_v \nabla^2 V \tag{13.40}$$

$$\frac{\partial I_u}{\partial t} = \alpha_{iu} F_1(X_1, C) - \gamma_{iu} I_u \tag{13.41}$$

$$\frac{\partial I_v}{\partial t} = \alpha_{iv} F_1(X_1, C) - \gamma_{iv} I_v \tag{13.42}$$

$$\frac{\partial C}{\partial t} = \alpha_c F_2(X_2, L) - \gamma_c C \tag{13.43}$$

where production rates of promoters $P_{Las-OR1}$ and $P_{Rhl-lacO}$ are $F_1(X_1, C) = \frac{[1+f_1(\frac{X_1}{K_{d1}})^{\theta_1}][1+f_2^{-1}(\frac{C}{K_{d2}})^{\theta_2}]}{[1+(\frac{X_1}{K_{d1}})^{\theta_1}][1+(\frac{C}{K_{d2}})^{\theta_2}]}$ and $F_2(X_2, L) = \frac{[1+f_3(\frac{X_2}{K_{d3}})^{\theta_3}][1+f_4^{-1}(\frac{L}{K_{d4}})^{\theta_4}]}{[1+(\frac{X_2}{K_{d3}})^{\theta_3}][1+(\frac{L}{K_{d4}})^{\theta_4}]}$, respectively. Las-A3OC12HSL complex and RhlR-IC4HSL complex are represented by $X_1$ and $X_2$, respectively. The concentration of unbound LacI protein is denoted by $L$. We define

$$X_1 = R_u U \tag{13.44}$$

$$X_2 = \frac{R_v V}{(1 + U/K_{c3})} \tag{13.45}$$

$$L = \lambda_l \left( \frac{1 + f_6^{-1}(I/K_{d6})^{\theta_6}}{1 + (I/K_{d6})^{\theta_6}} \right) \tag{13.46}$$

where $R_u$ and $R_v$ are regulatory LasR and RhlR regulatory proteins, respectively. The concentration of IPTG is $I$. We also define

$$R_u = \lambda_u I_u \tag{13.47}$$

$$R_v = \lambda_v \Big( \frac{1 + f_5^{-1}(C/K_{d5})^{\theta_5}}{1 + (C/K_{d5})^{\theta_5}} \Big) \tag{13.48}$$

where $\lambda_u$, $\lambda_v$, $\lambda_l$ and $K_{c3}$ are ratio between LasR and LasI, steady state level of RhlR, steady state level of LacI from expression $p_{lacq}$ and A3OC12HSL-RhlR dissociation constant, respectively. $\theta_1$, $\theta_2$, $\theta_3$, $\theta_4$, $\theta_5$ and $\theta_6$ are Hill coefficient for LasR-A3OC12HSL complex activation of $P_{Las-OR1}$, CI repression of $P_{Las-OR1}$, RhlR-IC4HSL complex activation of $P_{Las-OR1}$, LacI activation of $p_{Rhl-lacO}$, CI activation of $\lambda_{P(R-O1)}$ and IPTG binding to LacI, respectively. $K_{d1}$, $K_{d2}$, $K_{d3}$, $K_{d4}$, $K_{d5}$ and $K_{d6}$ represent dissociation constant of LasR-A3OC12HSL complex with $P_{Las-OR1}$, CI with $P_{Las-OR1}$, RhlR-IC4HSL complex with $P_{Las-OR1}$, LacI with $p_{Rhl-lacO}$, CI with $\lambda_{P(R-O1)}$ and IPTG with LacI, respectively. Furthermore, $f_1$, $f_2$, $f_3$, $f_4$, $f_5$ and $f_6$ are fold change for full induction of $P_{Las-OR1}$, inhibition of $P_{Las-OR1}$, induction of $P_{Las-OR1}$, inhibition of $p_{Rhl-lacO}$, induction of $\lambda_{P(R-O1)}$, induction of LacI activity for IPTG, respectively [17].

The results of deterministic model indicate that system is beyond the regime where classical Turing is formed. The volume of bacterial cells is small and many reactants are present in low numbers which are the basis reason behind the inherent noisiness gene expression in bacteria. So there is a possibility of finding stochastic Turing patterns in the system. Patterns are emerged in a parameter range which is expanded by the noise in stochastic Turing patterns. We construct a stochastic model for our system with same biochemical reactions, rate constant and diffusion as deterministic model. The motivation of this stochastic approach is to improve the correlation with experimental observations. We stimulate the stochastic model by tau-leaping stochastic algorithm. Stimulation of the model generates patterns with large variability in shape, size and intensity and intervals. This model generated patterns are in good aggregate with experimentally observed patterns [17]. We also perform power spectrum analysis. Noise behaves as a stabilizing agent in Turing like system. The patterns are formed for a wide range of parameters in the case of extrinsic and demographic noise [17]. This type of patterns is called stochastic Turing patterns. The study is a fundamental evidence of biological morphogenesis from theoretical and experimental aspects.

## References

1. Bashor CJ, Horwitz AA, Peisajovich SG, Lim WA (2010) Rewiring cells: synthetic biology as a tool to interrogate the organizational principles of living systems. Ann Rev Biophys 39:515–537
2. Endy D (2005) Foundations for engineering biology. Nature 438(7067):449–453

3. Khalil AS, Collins JJ (2010) Synthetic biology: applications come of age. Nat Rev Genet 11(5):367–379
4. Nandagopal N, Elowitz MB (2011) Synthetic biology: integrated gene circuits. Science 333(6047):1244–1248
5. Marguet P, Balagadde F, Tan C, You L (2007) Biology by design: reduction and synthesis of cellular components and behaviour. J R Soc Interface 4(15):607–623
6. Simpson ML (2004) Rewiring the cell: synthetic biology moves towards higher functional complexity. Trends Biotechnol 22(11):555–557
7. Majumdar S, Pal S (2018) Information transmission in microbial and fungal communication: from classical to quantum. J Cell Commun Signal 12(2):491–502
8. Gardner TS, Cantor CR, Collins JJ (2000) Construction of a genetic toggle switch in Escherichia coli. Nature 403(6767):339–342
9. Elowitz MB, Leibler S (2000) A synthetic oscillatory network of transcriptional regulators. Nature 403(6767):335–338
10. You L, Cox RS, Weiss R, Arnold FH (2004) Programmed population control by cell–cell communication and regulated killing. Nature 428(6985):868–871
11. Basu S, Gerchman Y, Collins CH, Arnold FH, Weiss R (2005) A synthetic multicellular system for programmed pattern formation. Nature 434(7037):1130–1134
12. Balagaddé FK, Song H, Ozaki J, Collins CH, Barnet M, Arnold FH, Quake SR, You L (2008) A synthetic Escherichia coli predator–prey ecosystem. Molecu Syst Biol 4(1):187
13. Danino T, Mondragón-Palomino O, Tsimring L, Hasty J (2010) A synchronized quorum of genetic clocks. Nature 463(7279):326–330
14. Lentini R, Martín NY, Forlin M, Belmonte L, Fontana J, Cornella M, Martini L, Tamburini S, Bentley WE, Jousson O, Mansy SS (2017) Two-way chemical communication between artificial and natural cells. ACS Cent Sci 3(2):117–123
15. Turing MA (1950) Computing machinery and intelligence. Mind 59(236):433–460
16. Turing A (1952) The chemical basis of morphogenesis. Philos Trans R Soc Lond Ser B Biol Sci 237:37–72
17. Karig D, Martini KM, Lu T, DeLateur NA, Goldenfeld N, Weiss R (2018) Stochastic Turing patterns in a synthetic bacterial population. Proc Natl Acad Sci 115(26):6572–6577

# Role of Noise in Synthetic Biology 14

**Abstract**

We consider noise as an inevitable part of living and non-living systems. We know synthetic biology is an interdisciplinary field. Engineering and basic science are patched up to understand complex biological phenomena. We discuss in detail about the conception of noise in synthetic biology from destructive point of view and/or engineer point of view as well as the constructive nature of noise and/or basic science point of view.

## 14.1 Introduction

Noise plays a destructive role in science and technology. But the discovery of stochastic resonance in nonlinear dynamics changed our idea regarding the role of noise in living organism which has been elaborated in the next chapter. Subsequently it raises an important issue how this change is envisaged in the context of synthetic biology. In fact, the fluctuations have been mentioned in the investigation of synthetic biological systems. In 2000 Elowitz and Leibler [1] introduced a gene oscillatory network known as repressiliator which exhibits fluctuations in its oscillations. It immediately triggers the question regarding the source of this noise and its interaction with the system. Different types of modelling approaches such as stochastic models, synthetic models and computer simulations were studied to respond these questions. Moreover, green fluorescent proteins based imaging technology were used for inspection of the dynamics on a single cell level. We have already mentioned that the investigations in nonlinear dynamics shifted the paradigm from the destructive nature of noise to constructive nature of noise which has deep impact in biological systems. But in which way the researches in synthetic biology changed the meaning of noise?

Firstly, the study of synthetic system repressilator brought in an empirical component. Recently, Loettgers [2] emphasized that synthetic systems could uncover



S. Majumdar, S. Roy, *Microbial Communication*,
https://doi.org/10.1007/978-981-15-7417-7_14

extra features with in vivo studies which gives us an opportunity to make noise visible with help of image technology. To get more insights, we need to discuss about the synthetic biology itself. One can discriminate two different branches of synthetic biology: one is basic science and another is application oriented. It is to be noted that most of the scientists working in second category, i.e. application oriented, have with engineering related background and those in the first category, i.e. basic science branch, have with physics related background. The scientists with their respective background approach the problem from their respective conceptual perspectives. For example, the engineers and physicists are expected to take different point of views to designing and exploring synthetic systems. Of course, there are some common grounds which can be shared by both the communities. Concepts like noise, feedback loops and robustness are considered as common grounds for both the groups.

It is important to note that sometimes developmental biologist say: *Why should we care about noise given the fact that biological organisms function despite the noise in and around it?* This seems to be an important argument. But once we consider the biological systems as organized mechanisms, evolved to deal with noise, noise has to be taken into consideration on how to use it. Subsequently, noise has become remarkable feature of biological system. Moreover, to understand the limits of synthetic biology in the engineering of new parts or systems the consideration of the contribution of noise is inevitable. We discuss the role of various types of noise in synthetic biology.

## 14.2 Noise in Synthetic Biology

In the beginning, synthetic biology helped us to update our understanding of why and when and how specific gene are expressed giving rise to new biological insights on noise or stochastic gene expression. Within isogenic bacterial populations gene expression varies from cell to cell even in steady state where the population reaches a stable average expression level. This is due to external as well as internal noise. The external noise results from the variations in the number of hardware units such as transcriptional–translational machinery and regulatory molecules, whereas the internal noise is associated with the random nature of single-molecule kinetics. These two types of noise play very important role in understanding the phenotype heterogeneity in genetically identical populations. The randomness associated with the noise is considered to be intrinsic property of gene expression which can be observed in artificial cells composed of cell membrane–mimetic vesicles containing only transcription and translation machinery [3].

We have already mentioned that there exist two approaches to synthetic biology; one from engineering background and the other mainly from physicist background. Subsequently, Knuutila et al. [4] studied the noise from these two approaches too. It is interesting to note that how the different scientific backgrounds fix the goals of research, various types of concepts from different fields, their interpretations, etc. For example, the notion of noise provides an important example of the similarities and differences between the above-mentioned two approaches. The synthetic biologists working in the engineering oriented approach treat noise as a destructive one which should be reduced to as small as possible, whereas those working in basic science-oriented approach concern the functional aspects of noise. In both the approaches, the meaning of noise retained as unwanted variation but the latter group address its functional role. They believe it to be a crucial and distinctive characteristic of biological processes. Again to address the functional aspect in biology Harttwell et al. in 1999 emphasize that the reductionist approach considered in physics cannot be accounted to one molecule but may be arise from the interactions of the components. To account for biological functions, they claim *we need a vocabulary that contains concepts such as amplification, adaptation, robustness, insulation, error correction and coincidence detection.* The key point in this discussion is that the physicists identify the relevant concepts from physics which are important for describing the biological functioning. Recent investigations on application oriented approach, it is now recognized that the noise also plays an important role in functional aspect of biological systems.

In engineering science, many types of noise appear but the entire filed in engineering is the study of reliable engineered systems. It raises an important issue how in case of engineered biological systems (like synthetic biology) the reliability be achieved ? It is important because of the existence of *ontic randomness* like genetic fluctuations which are intrinsic to biological systems. This type of randomness or fluctuations called noise makes it difficult to engineer reliable systems. It is to be noted that in application oriented branch of synthetic biology it is not our aim to mimic biological systems but to engineer novel systems with specific functions. Keeping this goal in mind and also because of the close ties with engineering, noise is considered as disturbance to the extent the control over the designed biological system is degraded. So the designing of a robust synthetic gene network which can tolerate intrinsic randomness and also function properly in host cells becomes one of the important topics of synthetic biology.

Chen and Li [5] made an interesting attempt to study the robustness of gene regulatory networks using concept of entropy. The concept of entropy is used in physics as a measure of degree of randomness. According to the second law of thermodynamics the entropy of a closed system increases with time. In principle, the entropy—a statistical law describes the most likely behaviour of the physical system. It immediately raises the question how this principle is applicable to living systems? The living matter seems to be a pretty ordered collection, every cell has its own internal organization, cells into tissues and tissues into organs, etc., i.e. animals, plants are all developed from an embryo. This orderliness appears to contradict the second law of thermodynamics. It is possible that some mechanism exists in biological system to counteract the increase of entropy. It is to be noted that biological systems are not closed and isolated systems in the sense of systems considered in the context of thermodynamic systems. They are usually considered as open systems interacting with the environment. This type of issue is widely discussed for open systems and the status of the principle of increase of entropy. The recent progress in synthetic biology indicates that noise contributes too many levels of gene and network regulation. They are studied through elegant synthetic circuit design.

Ciechonska et al. [6] raised a legitimate question in their recent article that can synthetic biology give us new fundamental and significant insights. Various disciplines like system biology, even classical cell biology are converging with synthetic biology to give rise to a new level of predictability to cellular signalling networks and gene expression, etc. The synthetic biology make significant contribution in comprehending gene expression with respect to noise or in other words in the view of cell heterogeneity. Here, they compared the progress in mammalian and bacterial systems which give us various most-developed engineering frameworks. The researchers in the last decade focus their attention on single cell biology. Probably, in coming decades, the advances into multicellular and synthetic biology will give rise to new therapeutic insights where noise may play very important role in contrast to conventional engineering domain. Ciechonska et al. envisaged an exciting change for synthetic biology as *"creating in order to understand" towards "creating in order to cure."*

## References

1. Elowitz MB, Leibler S (2000) A synthetic oscillatory network of transcriptional translation. Nature 403:335–338
2. Loettgers A (2010) Synthetic biology and the emergence of a dual meaning of noise. Biol Theory 4:340–356.
3. Nishimura K, Tsuru S, Suzuki H, Yomo T (2015) Stochasticity in gene expression in a cell-sized compartment. ACS Synth Biol 4(5):566–576
4. Knuuttila T, Loettgers A (2014) Varieties of noise: analogical reasoning in synthetic biology. Stud Hist Philos Sci A 48:76–88
5. Chen BS, Li CW (2010) On the interplay between entropy and robustness of gene regulatory networks. Entropy 12(5):1071–1101
6. Ciechonska M, Grob A, Isalan M (2016) From noise to synthetic nucleoli: can synthetic biology achieve new insights? Integr Biol 8(4):383–393

# Noise in Science and Technology vs. Biological System

# 15

**Abstract**

Noise plays an important role in science, engineering as well as in biological systems. The destructive role of noise has been extensively discussed in communication systems. The discovery of the phenomena called stochastic resonance clearly indicates the constructive role of noise for the nonlinear systems and hence in case of biological systems. Here in this chapter we discuss the important role of noise in cellular communication and sensory biology. Finally brief discussion is initiated on the concept of randomness and noise in science and engineering.

## 15.1 Introduction

> If the channel is noisy it is not in general possible to reconstruct the original message or the transmitted signal with certainly by any operation on the received signal. There are ways, however, of transmitting the information which are optimal in combating noise.—Claude E. Shannon [1].

In 1940, the concept of noise was introduced by Shannon and Weaver in the context of communication theory. Usually the sender sends a message and noise is something which is not included in the message. Here, one usually deals with the mechanical origin of noise, for example, noise distortion on the telephone or interference with the television signal which produces show on TV screen. The semantic noise arises from the ambiguities inherent in all languages and various sign systems.

It is noted that noise plays key role in communication of signal. In any physical system, the fundamental fact is that when the signal voltage arrives the demodulator, it will be accompanied by a voltage waveform which varies with time in an unpredictable manner. This unpredictable voltage waveform constitutes a random


S. Majumdar, S. Roy, *Microbial Communication*,
https://doi.org/10.1007/978-981-15-7417-7_15

process and usually known as noise. Here, the signal plays destructive role since it corrupts the signal.

## 15.2 Representation of Noise

Noise is generally passed through filters in communication system. The characteristics of these filters are usually described in frequency domain. So it is necessary to study the characteristics of the noise in the frequency domain to understand the effect of filters on the noise. Now we discuss such frequency domain characterization. This kind of representation will help us to define power spectral density for a noise waveform which has the similar characteristics to those of the power spectral density of a deterministic waveform. Taub and Schilling [2] made a comprehensive discussion for the representation of noise in frequency domain. They started with a noise sample function as $n^{(s)}(t)$. Let us consider this sample function of the noise and select an interval of duration $T$ extending, say, from $t = -\frac{T}{2}$ to $t = \frac{T}{2}$. Then it can be generalized considering a periodic waveform in which the waveform in the selected interval is repeated every $T$ s. The above-mentioned periodic waveform $n_T^{(s)}(t)$ can be expanded in a Fourier series and such a series represent $n^{(s)}$ in the interval $-\frac{T}{2}$ to $\frac{T}{2}$. Now power spectral density can be defined at the frequency $k\Delta f$ as the quantity

$$G_n(k, \Delta f) = G_n(-k, k\Delta f) = \frac{c_k^2}{4\Delta f} = \frac{a_k^2 + b_k^2}{4\Delta f} \tag{15.1}$$

where

$$c_k^2 = a_k^2 + b_k^2$$

So the total power associated with frequency interval $\Delta f$ is written as

$$P_k = 2G_n(k, \Delta f)\Delta f \tag{15.2}$$

One gets specific values of the coefficients $a_k$ and $b_k$ depending on the sample function of noise. If the representations as discussed earlier, are to represent random process, then these coefficients $a_k$ and $b_k$ are not fixed numbers but are random variables. It is to be noted that the power spectrum concept is useful because it helps us to resolve a deterministic waveform or a random process $f(t)$ into a sum

$$f(t) = f_1(t) + f_2(t) + \ldots. \tag{15.3}$$

in a manner by which the superposition of power applies, i.e. the power of $f(t)$ is the sum of the powers of $f_1(t), f_2(t), \ldots$.. In fact, a noise waveform can be represented as a superposition of spectral components, all of which are harmonics of some fundamental frequency. So far we talked about the representation of noise in frequency domain. The representation of noise is also possible in case of amplitude modulated systems [2].

Noise is usually considered as a truly fundamental engineering problem particularly in electronics computation and communication sciences, where the aim has been reliability optimization. The goal of communication engineers is to ensure transmission of error-free messages from one place to other by the fastest possible manner. So noise plays destructive role in signal analysis and communication system in science and engineering. The progress of researches on theoretical and experimental biological systems clearly indicates that the addition of input noise improves detectability and transduction of signals in nonlinear systems. Let us now discuss the role of noise in biological system.

## 15.3 Noise in Biological Systems

The discovery of Stochastic Resonances (SR) in nonlinear dynamics brought a shift to that perception, i.e. noise rather than representing a problem became a central parameter in system function, especially in biology. SR has been found to be an established phenomenon in sensory biology but it is not presently determined to what extent SR is embedded in such systems. Before going into the details of the application of SR to biological systems, let us briefly discuss the phenomena called Stochastic Resonance in nonlinear dynamics [3].

*In biology, "noise" typically refers to variability in measured data when identical experiments are repeated, or when bio-signals cannot be measured without background fluctuations distorting the desired measurement.* In some situations, unpredictable variations or unavoidably present fluctuations can be intentionally or deliberately introduced to get the benefit. Stochastic Resonance (SR) is such type of phenomena which is a flagship example of this idea [4, 5]. Benzi et al. [5] first introduced the concept of stochastic resonance in the context has been made in understanding the climate evolution. Physicists studied SR for more than 25 years and also by biologists, chemists and many from other disciplines. Benzi and his collaborators put forward this concept while they were addressing the problem related to periodically recurrent ice ages. Infact, Nicolis and Nicolis [6] independently discussed the role of stochastic resonance in the context of periodicity of the primary cycle of recurrent ice ages.

The first experimental verification of Stochastic Resonance was done by Fauve and Heslot [7] in 1983 by investigating the spectral line of ac-driven Schmitt trigger and its noise dependence. However, the experiment [8] in bistable ring laser triggered a large interest among the scientific community. Now it is necessary to define the observables so as to quantify the effect [9]. It is important to note that first of all the observable should be physically relevant and measurable. Benzi et al. quantified the stochastic resonance by the intensity of a peak in the power spectrum. In neurophysiology, another measure becomes popular, i.e. the interval distribution between activated events. For example, the interval distribution is given by successive firing of spikes of neurons. In 1991, Longtin et al. [10] found a remarkable resemblance between the residence-time and the inter-spike interval of periodically stimulated neurons and the residence-time distributions of periodically driven bistable systems. For convenience we discuss now the role of noise in two important areas of biology. At first, we discuss the relevance of stochastic resonance in sensory biology.

### 15.3.1 Stochastic Resonance in Sensory Biology

We have already mentioned that the term Stochastic Resonance was first introduced by Benzi et al. [4, 5] in climatology. Recently, Mark D. McDonnell1 and Derek [3] published an extensive review on the subject dedicated to biological systems. Broadly speaking, SR occurs in case of certain type of nonlinear systems where the stochastic noise enhances the coherence of the output subject to weak input signals. Here, the coherence is increased instead of its degradation. In early years, this effect was observed for the periodic input signals. Recently, this is also observed in case of random aperiodic input signals. Here, the information measure like Fisher information or mutual information is very useful one [11]. Viking rightly used the phrase *"noise benefits signal-processing systems rather than noise-enhances signal processing and so we declare SR as a noise benefit. But, does this variability really have a useful function? The study of noise enhanced phase synchronization of coupled oscillator arrays and nonidentical non-coupled noisy oscillators is a step ahead in understanding the benefits of noise in cell biology"*. The phase synchronization of noisy neuronal nonidentical oscillator was investigated by Neiman et al. [12] both from experimental observations as well as based on numerical simulation using stochastic synchronization. This type of synchronization has been investigated in different biological systems, for example, in case of fire flashing fireflies, in human cardio respiratory synchronization.

However, the issue which needs to be resolved whether SR occurs at the level of single ion-channel or an ensemble property of the aggregate of channels. It is possible that the intrinsic noise of the channel (known as internal noise) becomes ordered through a process known as intrinsic coherence resonance. This occurs even in the absence of external periodic signal. The provocating idea whether SR is exploited by the nervous system and brain as part of the neural code was put forward by McDonnell et al. The answer is "yes" and this has been recently demonstrated in the analysis of spontaneous oscillation of inferior olive cells following genetic mutation of particular ionic channel expression. In this case the absence of T or P type calcium channels results in a modification of coupled network oscillatory characteristics and in abnormal motor behaviour [13]. It is worth mentioning in our present discussion that the inferior olive cells in these mutants fail to generate the chaotic phase synchronization characteristic of this nuclear ensemble [14]. This lack of phase reset is rejected both in case of individual oscillation as well as in ensemble oscillations of neurons. Macharenko and Llinás demonstrated *"the electro-physiological characterization of the subthreshold oscillation in these mice demonstrated, in addition to the lack of phase reset, an asymmetry in subthreshold membrane potential oscillation"*. The neuronal variability was discussed by Stein et al. [15]. They raised a very important issue as to whether this variability is neural noise or a part of the signal transmitted to other neurones. The main argument put forward by them was both temporal and rate coding are used in various parts of Central Nervous System(CNS) and both are useful to CNS to discriminate complex objects and produce movements. The characteristics of noise in ion-channel is found to be of the nature of Flicker noise (FN), i.e. $\frac{1}{f}$. The mechanism of generation of FN in ion-channel is not yet fully understood. Now we go for another important area where noise plays also an important role. This is related to bacterial communication.

### 15.3.2 Noise in Cellular Communication

Inside the biofilms, bacteria coordinate their behaviour through different forms of communication mechanism than chemical signalling mechanism. Usually, cell-to-cell communication through chemical signalling molecules is popularly known as quorum sensing. Microbiologists intensively studied this critical biochemical phenomenon to understand the information processing system of different bacteria and their collective behaviour during last few decades. Bacterial communication system is controlled by autoinducers (chemical signalling molecules). Bacteria prepare their optimal survival strategies to survive in different environment by using different quorum sensing circuits. Quorum sensing bacteria eject autoinducers in the environment and the surrounding bacteria can take the autoinducers. In this fashion the concentration of the autoinducers increases as a function of cell number density. When the concentration reaches minimal threshold, a collective bacterial behaviour is initiated and this triggers cascade of signalling events and regulate an array of biochemical process such as biofilm formation, swarming, virulence, bioluminescence, symbiosis, competence, antibiotic production, sporulation, conjugation and gene expression [16]. Apart from the communication based on chemical signalling molecules bacteria communicate through potassium ion-channels mediated electrical signalling process and coordinate metabolism within the biofilm and hence conduct a long range electrical signalling within bacterial biofilm communities through the propagating wave of potassium [17, 18]. Motile bacterial cells are attracted towards biofilm and the attraction depends on membrane potential. Bacterial biofilms are in general bacterial community, which undergo metabolic oscillations and coupled through electrical signalling process ($K+$ waves) and synchronized biofilm growth dynamics. This coupling increases competition by also synchronizing demand for limited nutrients. It has been observed that biofilms resolve this conflict by switching from in phase to anti-phase. Different biofilm communities take turns consuming nutrients. Thus, distant biofilms can coordinate their behaviour to resolve nutrient competition through time-sharing [19].

One of the present authors (SR) along with Rodolfo Llinas [20] emphasize that densely packed bacteria may be viewed as "bacterial fluid" or "living fluid" similar to that of dense granular systems. In this framework, there exists a non-thermal fluctuation associated with the finite size of the bacteria similar to the finite size of the granules, which is called non-local noise. This is non-thermal in nature. This kind of fluctuation helps to understand the communication of bacteria at the level where communication occurs through chemical signalling depending on the kinematic viscosity associated with this non-local noise. Complex Ginzburg–Landau equation is used to describe the behaviour of bacteria based on chemical signalling where the kinematic viscosity can be traced back to its origin to the above-mentioned non-local noise. The kinematic viscosity of bacteria will be small as it requires to have large density for quorum, so that emergence of metastable state is possible for short time scale. During this time period the fluctuation of stress due to autoinducing molecules will produce fluctuation in the configuration of the system which induce shear somewhere else. This process is known as self-activated process which occurs in rheology. This mechanical stress is responsible for the gene expression of the bacteria and quorum happens. It is evident from the above discussions that noise and randomness are closely related concepts. We first discuss noise and randomness in science and engineering. Then we discuss these concepts in the context of biological systems.

## 15.4 Noise and Randomness in Science and Engineering

Probably, the concept of randomness was first discussed by the philosophers from the East many centuries before Epicurus (341–270 BC). They thought it in the sense of unpredictability as related to manifestation of the universe. Epicurus argued that *randomness is objective, it is the proper nature of events*. Poincare made a major contribution towards the contemporary understanding of randomness. The term chance is used for many centuries in relation to many human activates like gambling, etc. This is essentially related to the lack of knowledge and this lack of knowledge in human activities are estimated based on the toll called probability. Subsequently, the connection between randomness and incomplete knowledge of natural phenomena is established. Then the formal calculus of randomness is constructed through probability theory of course with no commitment to the nature of randomness. Before the birth of quantum theory, the form of randomness is considered to be "epistemic" unpredictability, i.e. as related to our lack of knowledge of the world. However, the randomness in the phenomena called Brownian motion in classical physics is an exception and it is ontic or intrinsic in nature.

Quantum randomness and quantum chance are considered to be more than epistemic that is "intrinsic". Classical randomness in contrast to quantum randomness is generally used in the field of game theory, random motion of molecules, etc. Random processes have been extensively studied in probability theory, ergodic theory and information theory.

Information theory has been extensively studied in science and engineering due to pioneering work of Shannon. Here, the terms randomness and noise have been used in understanding signal processing in communications. Noise is generally consider as a truly fundamental problem in engineering, electronics and communication process, where the aim has been reliability optimization. Engineers normally use signal processing techniques to distinguish between useful signals in communication instruments and interference which disturbs the desired signal. The knowledge of fundamentals of stochastic processes and their practical applications help to understand the random signals and noise.

## 15.5 Randomness in Biology

In 2006 Arias and Hayward [21] made a relevant statement *"The large number of cell states that are present in the lifetime of an organism and the reproducibility with which they are generated indicates the existence not just of programmes but also of mechanisms that ensure their reliable execution"*. This is considered to be the dominant belief of life sciences. In this circumstances it might be surprising to talk about the randomness in biology. It is well known that chance has been dealt in the long history of life sciences. In 1827 English botanist Robert Brown who happened to be early sources of inspiration of Darwin discovered the zigzag movement of pollen grains in water popularly known as Brownian movement. The role of any biological force as an explanation for this movement was rejected because this was observed in case of inorganic situation. Almost a century later (1905) Albert Einstein gave its theoretical explanation based on the law of physics. Here, Einstein introduced concept of randomness which is intrinsic hence ontic in nature. In many ways, randomness is considered as intrinsic feature of evolutionary biology and genetics. In contrast to evolutionary biology and genetics, deterministic framework is shown to be suitable for understanding molecular biology. The recent theoretical and experimental results have challenged this view. They provide unifying explanations based on intrinsic stochastic dimension rather than just as background noise [22].

## References


1. Shannon CE (1948) A mathematical theory of communication. Bell Syst Tech J 27(3):379–423
2. Taub H, Schilling DL (1986) Principles of communication systems. McGraw-Hill Higher Education, New York
3. McDonnell MD, Abbott D (2009) What is stochastic resonance? Definitions, misconceptions, debates, and its relevance to biology. PLoS Comput Biol 5(5):e1000348
4. Benzi R, Sutera A, Vulpiani A (1981) The mechanism of stochastic resonance. J Phys A Math Gen 14(11):L453–L457
5. Benzi R, Parisi G, Sutera A, Vulpiani A (1982) Stochastic resonance in climatic change. Tellus 34(1):10–16
6. Nicolis C, Nicolis G (1981) Stochastic aspects of climatic transitions–additive fluctuations. Tellus 33(3):225–234
7. Fauve S, Heslot F (1983) Stochastic resonance in a bistable system. Phys Lett A 97(1–2):5–7
8. McNamara B, Wiesenfeld K, Roy R (1988) Observation of stochastic resonance in a ring laser. Phys Rev Lett 60(25):2626

9. Gammaitoni L, Hänggi P, Jung P, Marchesoni F (1998) Stochastic resonance. Rev Modern Phys 70(1):223
10. Longtin A, Bulsara A, Moss F (1991) Time-interval sequences in bistable systems and the noise-induced transmission of information by sensory neurons. Phys Rev Lett 67(5):656
11. Kosko B (2006) Noise. Penguin, New York
12. Neiman A, Silchenko A, Anishchenko V, Schimansky-Geier L (1998) Stochastic resonance: noise-enhanced phase coherence. Phys Rev E 58(6):7118
13. Choi S, Yu E, Kim D, Urbano FJ, Makarenko V, Shin HS, Llinás RR (2010) Subthreshold membrane potential oscillations in inferior olive neurons are dynamically regulated by P/Q- and T-type calcium channels: a study in mutant mice. J Physiol 588(16):3031–3043
14. Makarenko V, Llinás R (1998) Experimentally determined chaotic phase synchronization in a neuronal system. Proc Natl Acad Sci 95(26):15747–15752
15. Stein RB, Gossen ER, Jones KE (2005) Neuronal variability: noise or part of the signal? Nat Rev Neurosci 6(5):389
16. Majumdar S, Pal S (2018) Information transmission in microbial and fungal communication: from classical to quantum. J Cell Commun Signal 12(2):491–502
17. Liu J, Prindle A, Humphries J, Sagarra MG, Asally M, Dong-yeon LD, Ly S, Ojalvo GJ, Süel MG (2015) Metabolic co-dependence gives rise to collective oscillations within biofilms. Nature 523(7562):550–554
18. Prindle A, Liu J, Asally M, Ly S, Ojalvo GJ, Süel MG (2015) Ion channels enable electrical communication in bacterial communities. Nature 527(7576):59–63
19. Liu J, Corral MR, Prindle A, Dong-yeon LD, Larkin J, Sagarra GM, Ojalvo GJ, Süel MG (2017) Coupling between distant biofilms and emergence of nutrient time-sharing. Science 356(6338):638–642
20. Roy S, Llinás R (2016) Non-local hydrodynamics of swimming bacteria and self-activated process. In: BIOMAT 2015: International symposium on mathematical and computational biology, pp 153–165
21. Arias AM, Hayward P (2006) Filtering transcriptional noise during development: concepts and mechanisms. Nat Rev Genet 7(1):34
22. Heams T (2014) Randomness in biology. Math Struct Comput Sci 24(3):e240308

# Index




S. Majumdar, S. Roy, *Microbial Communication*,
https://doi.org/10.1007/978-981-15-7417-7

The manufacturer's authorised representative in the EU is Springer Nature Customer Service Centre GmbH, Europaplatz 3, 69115 Heidelberg, Germany. If you have any concerns regarding our products, please contact ProductSafety@springernature.com

Printed and bound by CPI Group (UK) Ltd, Croydon, CR0 4YY
06/10/2026
02228179-0001